WRITING FOR TODAY'S HEALTHCARE AUDIENCES

WRITING FOR TODAY'S HEALTHCARE AUDIENCES

Robert J. Bonk

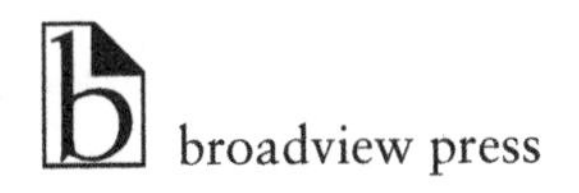

LIBRARY AND ARCHIVES CANADA CATALOGUING IN PUBLICATION

Bonk, Robert J., author
Writing for today's healthcare audiences / Robert J. Bonk.

Includes bibliographical references and index.
ISBN 978-1-55481-149-6 (pbk.)

1. Medical writing. I. Title.

R119.B65 2015 808.06'661 C2014-908075-1

BROADVIEW PRESS is an independent, international publishing house, incorporated in 1985.

We welcome comments and suggestions regarding any aspect of our publications—please feel free to contact us at the addresses below or at broadview@broadviewpress.com.

NORTH AMERICA
Post Office Box 1243
Peterborough, Ontario
K9J 7H5 Canada

555 Riverwalk Parkway
Tonawanda, NY 14150, USA
TEL: (705) 743-8990
FAX: (705) 743-8353

customerservice@broadviewpress.com

UK, EUROPE, CENTRAL ASIA, MIDDLE EAST, AFRICA, INDIA, AND SOUTHEAST ASIA
Eurospan Group, 3 Henrietta St., London WC2E 8LU, United Kingdom
TEL: 44 (0) 1767 604972 FAX: 44 (0) 1767 601640
eurospan@turpin-distribution.com

AUSTRALIA AND NEW ZEALAND
Footprint Books
1/6a Prosperity Parade
Warriewood, NSW 2102, Australia
TEL: 61 1300 260090
FAX: 61 02 9997 3185
info@footprint.com.au

www.broadviewpress.com

Broadview Press acknowledges the financial support of the Government of Canada through the Canada Book Fund for our publishing activities.

Edited by Martin R. Boyne
Typesetting by Em Dash Design

Printed in Canada

CONTENTS

LIST OF FIGURES AND TABLES

FIGURES

TABLES

ACKNOWLEDGMENTS

As a writer, finding the time to write this—or any—book is challenging. I therefore thank Widener University for granting me a one-semester sabbatical to provide additional time.

As an educator, I am especially pleased to watch my students grow into writers. This growth was especially apparent in one of my best students, Chara Kramer, who graduated in May 2014. During her last semester, she worked closely with me, through a practicum class, in reviewing and editing the nearly final chapters. I foresee success for Chara as a writer and editor.

As a person, I have needed to juggle many priorities while completing this book. I thank my family and friends for their support.

PREFACE

In our fast-paced technological society, writers must be able to communicate clearly and concisely with colleagues, associates, and clients; the same holds true for the world of healthcare. After all, without communication, healthcare advances would not improve the lives of others. Knowing how to write for these increasingly diverse users of healthcare audiences is the focus of this book.

If you're reading this book now, you're most likely a student or a professional new to writing about healthcare. You may be interested in any healthcare field—from pharmacy, nursing, and other specialties to the nascent yet growing discipline of medical writing. This book will help you to develop your personal abilities to write efficiently and effectively for various healthcare audiences.

Together, we'll explore today's healthcare writing with just enough theory to lay groundwork, plentiful examples to illustrate how this theory is practiced, summaries that highlight key points, and realistic exercises to practice what you're learning. Even though we're focusing on audience, you'll be introduced to the typical documents encountered in healthcare communication.

So let's begin now.

1

PRELIMINARIES OF HEALTHCARE WRITING

Chapter Objectives

In this chapter, you will learn to:

- Understand that effective writing begins with solid preparation
- Identify the input conditions of audience, purpose, and context
- Relate the output parameters of medium, content, and strategy
- Realize that healthcare writing has relevant ethical dimensions

Thinking as Readers, Not Writers

Imagine yourself being in a foreign country where you don't speak the local language, and the local people don't speak yours. To make matters worse, what if you were injured or needed some sort of healthcare? How would you make your needs known to doctors and nurses in the local hospital? Even if you did, how would they inform you of your medical status, not to mention instruct you on how to treat your problem? How would you have found that hospital at all? Assuming that these vital communications somehow occurred through diagrams and gestures, just imagine walking into the billing office!

Fortunately, few of us should ever find ourselves in the situation just described. This scenario, though, does illustrate a key point about writing in today's complex world: We who communicate healthcare information must learn to think *less* as ourselves who possess the knowledge, and *more* as those who need that knowledge. In other words, we must plan for effective communication that targets particular audiences. That planning initiates the dynamic process of writing.

Overview of the Writing Process

The writing process is both dynamic and simple. First, the process is dynamic because writers critically analyze and astutely adjust the flow of information; the process itself contributes to a more solid understanding of the topic. Second, the simplicity of the writing process belies the astuteness of its steps; careful effort in the early stages increases the efficiency of the overall process.

Essentially, writing as a process proceeds through three main phases:

1. *Prewriting*, when you plan your document and collect your information
2. *Writing*, when you draft your document and revise it for major changes
3. *Postwriting*, when you hone, polish, and finalize your completed document

Experienced writers have learned that the key to writing is emphasizing the first phase of prewriting. In fact, if you first carefully analyze your audience, you can then determine the level and types of information needed. For example, will readers expect statistics and literature references? Or will such complex information simply overwhelm your reader? Fundamental to this efficient process of writing is understanding how conditions of the healthcare situation determine parameters of the final document. In other words, the writing process has inputs that determine outputs.

Inputs and Outputs of Healthcare Writing

Audience + purpose + context = medium + content + strategy

This equation summarizes the fundamental elements of all types of professional writing—including communication pieces on healthcare topics. The left side of the equation identifies three key input conditions (*audience, purpose, context*) describing a writing assignment, whereas the right side identifies three key output parameters (*medium, content, strategy*) defining the resultant piece. Let's see how conditions on the left lead to parameters on the right. In other words, how do healthcare situations clue writers into crafting communication pieces that work?

The *audience* for healthcare information has diversified today: The traditional audiences of patients and physicians now encompass practitioners such as pharmacists, nurses, and therapists, as well as policy analysts, plan administrators, and others involved with the healthcare system. Healthcare documents have a *purpose* to achieve—to inform someone about a new surgery but not instruct that person how to perform it, for example. *Context*, the setting in which communication occurs, can be pivotal, especially when we consider emotionally charged emergency scenarios or end-of-life decisions.

These three input conditions identify for writers the three output parameters that inform them about how to construct the most suitable communication piece for that topic. First, the message is transferred through a chosen *medium*, be that a lengthy tome, a trifold pamphlet, or (increasingly) a web-based form of communication. The types and levels of information signal *content*, whereas *strategy* refers to the techniques used to organize and present the selected information.

Perhaps the opening mathematical equation should be expressed as a biochemical reaction: understanding the conditions (*audience, purpose, context*) points to determining the parameters (*medium, content, strategy*) for the resultant healthcare document. Or, more appropriately for this field, we are dealing with a diagnosis of the writing situation.

Typical Documents for Healthcare Writers

Embarking on new careers, healthcare writers (and students of healthcare writing) may encounter a wide variety of situations requiring communication pieces. Analyzing the conditions of these situations, as just explained, leads to determining the parameters for constructing the documents. Note that the term *document* in this sense loosely means some piece of communication: paper or online, short or long, print or audiovisual, and whatever new media may arise.

Later chapters explicate the typical healthcare documents, along with guidance and examples. The following non-exhaustive list offers a taste of what's to come:

- *Protocol*—a plan for conducting medical research (think of it as being like a procedure for a laboratory experiment in a science course)
- *Education Piece*—material intended to provide sufficient information for readers needing that level of detail (short or long textbooks, in essence)
- *Training Guide*—stepwise instructions for those who will actually perform that specific activity (oriented more for work than study)

- *Grant Proposal*—formal application to a governmental, corporate, or charitable organization for funding (persuasive case built from evidence)
- *Advocacy Piece*—persuasive argument on behalf of a patient group or medical condition (public health campaigns, as one example)

Other documents fall more into the researcher's realm: annotated bibliographies, literature reviews, and technical reports, for example. Although not covered in detail (as the researcher audience is not this book's focus), these document types are mentioned in relation to items in specific chapters.

Ethical Responsibilities of Healthcare Writers

Social Contract of Professionals

All professionals are responsible to the individuals whom they serve. Contrary to loose modern usage, a "professional" role is one that is needed by society but that requires a specialized set of skills. Think of doctors, ministers, and lawyers. Because these roles mandate a specialized set of knowledge, professionals and society reciprocally enter a "social contract"—to balance the professionals' expertise, society grants them authority to conduct services on society's behalf. Key to this social contract, however, is that services provided by the professionals must always remain in the best interests of society and society's members.[1]

How do healthcare writers fit into this social contract? Healthcare writers provide a unique service that not everyone in society can: translating complex research information into forms understood by layperson, administrator, and practitioner audiences. With that recognition of professionals come responsibilities to fulfill the contract. The American Medical Writers Association (AMWA) encapsulates this ethical responsibility as the onus to "apply objectivity, scientific accuracy and rigor, and fair balance while conveying pertinent information in all media" and to "write, edit, or participate in the development of information that meets the highest professional standards."[2]

Ethics, Respect, and Sensitivity

When actively working on a healthcare document, writers fully immerse themselves into a complex task. As explained in the next chapter, writers often need to make calculated assumptions about their target audiences; these decisions depend not only on pertinent information but also on respect for individuals and recognition of ethical dimensions. Unlike a user manual for a digital recorder, a self-assessment checklist for testicular cancer could place a man's health and life in jeopardy if the instructions are vague. That endangerment is much more serious, of course, than not recording the next installment of a favorite television show.

Remember, too, that calculated assumptions about a target audience are guidelines rather than absolutes: within the target group, individuals remain persons. Some men may be sensitive to candid wording about the testicular exam; explicit photos rather than stylized diagrams provide more details but may offend men from conservative cultures or religions; and locker-room humor could assuage trepidations for teens but alienate senior citizens. Many such issues may be obviated by thorough analysis in the prewriting stage, if the writer also considers respect for multicultural diversity.

A good rule of thumb is to run potentially sensitive issues past advocacy groups or support organizations. Would most members of a certain multicultural identity prefer one demographic term (such as *Latino*) over another (like *Hispanic*)? Moreover, respect is shown when recognizing the humanity of an ill person: a woman with gynecological problems, after all, is not *the hysterectomy down the hall.* An especially relevant issue in healthcare concerns persons with disabilities. The term *handicapped,* for instance, brings more negative connotations than *disabled*; more positive connotations may be associated with *differently abled*—or is that too politically correct?

Later chapters will return to issues such as these through specific examples and writing guidance. For now, remain aware of respecting the individuality of patients, caregivers, and family members. How would *you* feel in that situation? Also, take seriously the ethical dimensions of healthcare writing. The life of an individual may very well be at stake. The painstaking effort required for effective healthcare documents reaps rewards for the target audience.

Healthcare Writers as Advocates

As professionals, healthcare writers must work for the benefit of individuals who seek medical care. Much of this responsibility can be accomplished through diligently ensuring the accuracy and reliability of documents. Overt bias should be avoided at all times, even though some documents (e.g., materials from a drug company) almost always have their own slant. As long as information is presented clearly and fairly, the unavoidable instances of slant should not pose concerns. After all, even documents from a non-profit agency still reflect that group's agenda in terms of how they view and promote certain health issues.

Sometimes, however, the writer's position is more tenuous. Should writers take issue with a stance with which they do not fully agree? In many cases, writers work as full-time employees or contracted freelancers. While writers should be willing to raise "red flags" whenever undue biases or misrepresentations seem to be occurring, the final decision remains with the person or group in charge of the project. Only in rare situations will documents take unethical turns—at which times writers must be willing to stand up, even at financial risk. Professionals serve society.

Finally, some healthcare writers may find themselves in advocacy positions. For instance, after working on educational materials for a disadvantaged or disenfranchised group, writers may feel responsible for acting on that group's behalf. Reproductive rights provide such a touchstone. Should writers work on documents that, while unbiased, still conflict with their own ethical or religious viewpoints? For example, a media toolkit advocating for the availability of contraceptives, which has been produced under the aegis of the Johns Hopkins University, is obviously and understandably slanted in favor of contraceptive rights. Interestingly, that same toolkit contains excellent guidance for using the media for other advocacy issues (Figure 1.1).[3] Should a writer who is opposed to contraceptive use still take advantage of this media guidance?

Figure 1.1 Media Toolkit on Advocating for Contraceptives

V Media Targeting and Assessment

No media advocacy campaign can be effective without first discerning what mediums your target audience most often consumes and which media outlets they use the most. Whom are you trying to reach and what do they watch, read or listen to? Outlets should be ranked in terms of priorities. This simple process can help you to ensure that your efforts are dedicated to those outlets that are the greatest priority.

- What is the medium (television, radio, newspapers, magazines, Web) most often consumed by your targeted population? And which outlets are most used by your targeted population?
- Which are most popular with the general public? Among those media outlets, whom or what should you contact to promote your agenda? A particular reporter, editor or producer? A particular columnist? TV news show? A particular newspaper? A radio station or show?
- Do you know those people or have you had previous contact with them? Do you have their contact information?
- What media outlets and sources influence members of the media? This will be an important forum.

After you have determined which outlets are of the greatest importance to you, you should conduct some basic research to help you determine what obstacles you might face or what support can be amplified.

- Has the media covered contraceptive security/FP/contraception recently? Has it been supportive or hostile?
- Who covered the issue?
- If the media hasn't covered the issue, has it covered a similar health or women's issue?
- What reporter covered this issue?
- What other campaigns have been successful in engaging the media recently? What can you learn from them? What did they do and what was the result?
- Who does the media generally find compelling as a spokesperson?
- What can you practically apply to your campaign?

MAGAZINES
NEWS PAPERS
RADIO
TV

Lesson
Determine the media outlets and sources most important to your target audience. These outlets are the ones that you will try hardest to influence.

Media Advocacy Tool • 9

A book like this one cannot answer such ethical dilemmas. Rather, this book aims to increase awareness of our responsibilities as professionals. As appropriate, issues with ethical dimensions will be threaded throughout the book's examples, and that ethical thread is revisited in the final chapter.

Chapter Summary

- Emphasizing the prewriting stage allows healthcare writers to apply their time and effort efficiently by avoiding late-stage rewrites.
- Input conditions (*audience, purpose, context*) determine output parameters (*medium, content, strategy*) for healthcare documents.
- The same healthcare topic can be presented in various document types, depending on the input conditions for a particular situation.
- Although calculated assumptions can facilitate the writing process, writers always must remember that individuals must be respected.
- Because healthcare decisions can truly make the difference between life and death, healthcare writers must act within ethical dimensions.

Exercises for Practice

The exercises in this chapter refer to Figure 1.2 (p. 25), a document excerpt that includes both text and a visual related to the testicular self-examination (TSE).[4] Additional material accessible through the hyperlink includes a text version in Spanish and an audio version. Considering these materials, respond to the following questions.

1. How specifically can you describe the target audience—of course, they are men, but what other details can you surmise? What is the purpose of this document? Does the context make a difference? Would you suggest any modifications?

2. TSE information could appear in a number of different formats. Select at least two different types. Compare and contrast them for the input conditions of audience, purpose, and context. List ideas for the output parameters of medium, content, and strategy. How do your planned documents differ from the original excerpt?

3. Sensitive materials, such as TSE guidelines, can be uncomfortable. How explicit should the material be? What approaches can be taken to interest but not offend the target audience? Consider these questions as you assess the excerpt.

4. The additional excerpt of text (below) originates from the same document as the main excerpt in Figure 1.2. Why was this additional text written differently from the main excerpt? What are these differences?

> Although testicular cancer is rare in teenage guys, overall it is the most common cancer in males between the ages of 15 and 35. It's important to try to do a TSE every month so you can become familiar with the normal size and shape of your testicles, making it easier to tell if something feels different or abnormal in the future.... Lumps or swelling may not be cancer, but they should be checked by your doctor as soon as possible. Testicular cancer is almost always curable if it is caught and treated early.

5. Despite communication about the TSE, too many men still face testicular cancer. Support is important at these vulnerable times. Groups like the American Cancer Society provide support opportunities, such as through the online WhatNext social network.[5] How can such support groups reach out to individuals perhaps more effectively than through formal documents?

Figure 1.2 Text Excerpt with Visual for the Testicular Self-Examination (TSE)

How to Perform a Testicular Self-Examination

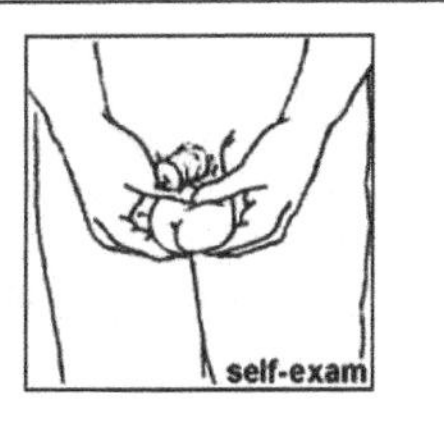

The **testicular self-examination (TSE)** is an easy way for guys to check their own testicles to make sure there aren't any unusual lumps or bumps—which can be the first sign of testicular cancer.

Here's what to do:

- It's best to do a TSE during or right after a hot shower or bath. The scrotum (skin that covers the testicles) is most relaxed then, which makes it easier to examine the testicles.
- Examine one testicle at a time. Use both hands to gently roll each testicle (with slight pressure) between your fingers. Place your thumbs over the top of your testicle, with the index and middle fingers of each hand behind the testicle, and then roll it between your fingers.
- You should be able to feel the epididymis (the sperm-carrying tube), which feels soft, rope-like, and slightly tender to pressure, and is located at the top of the back part of each testicle. This is a normal lump.
- Remember that one testicle (usually the right one) is slightly larger than the other for most guys—this is also normal.
- When examining each testicle, feel for any lumps or bumps along the front or sides. Lumps may be as small as a piece of rice or a pea.
- If you notice any swelling, lumps, or changes in the size or color of a testicle, or if you have any pain or achy areas in your groin, let your doctor know right away.

Lumps or swelling may not be cancer, but they should be checked by your doctor as soon as possible. Testicular cancer is almost always curable if it is caught and treated early.

References

1 Bonk RJ. Societal framework of health and medicine: recommendations for medical writers. *Am Med Writers Assoc J.* 2004;19(1):12-15. http://www.amwa.org/issues_online.

2 American Medical Writers Association. *Code of Ethics.* 3rd rev. Rockville, MD: American Medical Writers Association; June 2008. http://www.amwa.org/amwa_ethics.

3 Cuttino P, Negrett JC. *Media Advocacy for Contraceptive Security: A Tool for Strategy Development.* Baltimore: The Health Communications Partnership based at The John Hopkins Bloomberg School of Public Health / Center for Communication Programs; September 2006. https://www.jhuccp.org/resource_center/publications.

4 Figueroa TE, reviewer. *How to Perform a Testicular Self-Examination.* Wilmington, DE: TeensHealth, reviewed June 2012. http://teenshealth.org/teen/sexual_health/guys/tse.html.

5 WhatNext. Atlanta: The American Cancer Society; 2014. https://www.whatnext.com.

2

AUDIENCES FOR HEALTHCARE INFORMATION

Chapter Objectives

In this chapter, you will learn to:

- Distinguish primary vs. secondary audiences for documents
- Categorize healthcare audiences by their level of knowledge
- Gain an overview of different techniques for audience types

Primary and Secondary Audiences

Among the three conditions that serve as inputs in healthcare writing, one provides the most information—audience. Taking sufficient time during the prewriting stage allows writers to clarify as many descriptive details and calculated assumptions as possible about those who need access to the healthcare topic. Let's get to know our audiences.

On a broad level, we can distinguish two main groups: *primary* and *secondary*. Those who are intended to use the information form the *primary* audience. For example, patients in a clinic would be the audience for a brochure describing how to recognize the telltale symptoms of diabetes. In contrast, a medical technologist would be the primary audience for a manual on operating an instrument for measuring glucose levels in blood serum. Clearly, you would need to write these two documents quite differently—and you would plan for those differences in the prewriting stage. Giving the same information to both audiences in the same way will not suffice.

Besides having primary audiences, some documents may also have a *secondary* audience—someone else who may want to access that same information. In the case of the patient's brochure on glucose, the doctor and nurse in the clinic would need to know what information is covered so that they can anticipate any questions from the patients. For the manual, the researchers who designed the instrument would want to ensure that all information is correct; technologists may review the manual for clarity; lawyers may even need to review it for any safety liabilities. (Some writing texts consider the lawyers in this example as tertiary because of their displacement from the topic.)

Fortunately, in many documents, a secondary audience can be accommodated easily by adding material in an appendix (e.g., operating principles for the laboratory equipment) or sometimes by stating a disclaimer (e.g., liability if safety warnings are ignored). When a document has a secondary audience, always try to address the concerns and interests of these other readers. However, never place the preferences of the secondary audience over the needs of the primary audience. If necessary, consider writing more than one type of document so that all relevant audiences are addressed.

Categorization by Knowledge Level

Knowing the importance of writing for the primary audience is a start, but that term remains somewhat broad and potentially vague at this point of prewriting. How do writers identify the primary audience's needs, wants, and preferences for the document? To do so, writers must delve deeper into the minds of potential readers; this approach is analogous to the way in which marketing companies collect consumer profiles.

One useful technique for understanding readers is to categorize the audience by their level of knowledge about the topic. Knowledge level, though, specifically refers to the healthcare topic, not to the general educational level of the audience.[1] A judge certainly has a high knowledge level about jurisprudence, further substantiated by her degrees, but that same judge may need the same reassurance as other patients when facing breast cancer. Knowing even more about the reader's knowledge may even point toward unexpected approaches for a document: an auto mechanic may not fully comprehend a dense journal publication on congestive heart failure, but he would certainly relate to a description and diagram patterned on a car engine.

Categorizing readers by knowledge of the healthcare topic leads to four principal groups: *layperson*, *administrator*, *practitioner*, and *researcher* (Figure 2.1). These groups differ by not only their knowledge level but also their reasons for reading. *Laypersons* typically have just a cursory familiarity with a healthcare topic (although a wide public audience of mainly laypersons will also encompass individuals from the other audience groups). Generally, laypersons read from personal interest, perhaps about a therapy for a family member with an illness. *Administrators,* the next group, can be difficult to place. Some administrators, such as insurance agents or pharmaceutical representatives, may have business degrees supplemented by on-the-job training; others may be physicians, pharmacists, or nurses with dual roles as business managers. Regardless, individuals in this category do not read from personal interest but to make key decisions on economic resources such as time and money.

Figure 2.1 Categories of Healthcare Audiences by Knowledge Level

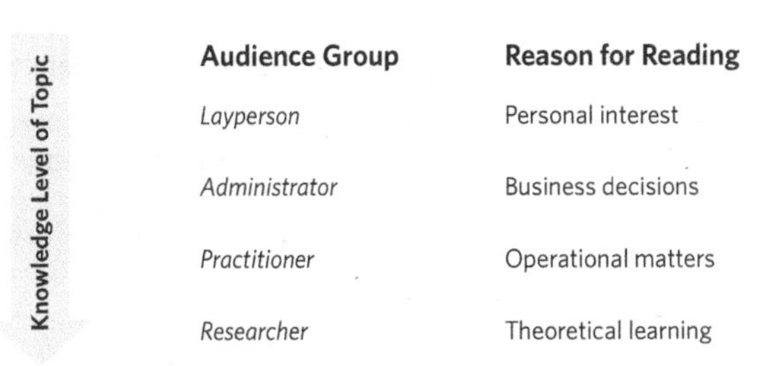

Knowledge Level of Topic
Audience Group
Reason for Reading
Layperson
Personal interest
Administrator
Business decisions
Practitioner
Operational matters
Researcher
Theoretical learning

The next two groups, *practitioners* and *researchers,* may be equivalent in terms of knowledge. They differ, though, in their motivations for using the information: practitioners want to carry out activities operationally, whereas researchers are more interested in the theoretical underpinnings. To clarify, consider practitioners as primary-care physicians, pharmacists, or medical technologists diagnosing and treating a child's cancer; think of researchers as medical and nursing professors at a university specializing in pediatric oncology. Nonetheless, be aware that individuals can fall into both categories. A nurse caring for a hospital ward would be a practitioner; he may also be a researcher testing different strategies for quality of care.

In today's healthcare environment, these categories become even more complex. Take the United States, a nation witnessing a proliferation of healthcare roles, particularly for medical specialties and insurance plans (Table 2.1). Patients, who are part of the layperson group, are increasingly diverse, not just in terms of sociocultural factors but also with respect to literacy levels. And the other three groups are just as varied: administrators include sales and marketing staff, insurance agents, and government policy makers; practitioners apply knowledge for each new health specialty; and researchers fill the niches revealed by biotechnology. As all countries grapple with providing healthcare outcomes at a feasible cost, approaches for understanding healthcare audiences beyond the broad level of knowledge will be crucial.

Table 2.1 Proliferation of Healthcare Roles for Audience Categories

	AUDIENCE CATEGORY	
Healthcare Roles	Layperson	Administrator
Traditional	Adults with limited healthcare knowledge	Sales and marketing staff of drug industry
Expanding	Educated consumers Semi-literate readers Non-native speakers Discrete age groups	Insurance agents Plan administrators Government officials Policy analysts

Table 2.2 Example Documents for Categories of Healthcare Audiences

	AUDIENCE CATEGORY	
Writing Techniques	Layperson	Administrator
Overall focus	Personal interest and human appeal	Emphasis on money and other resources
General language	Informal tone, often with contractions	Content that allows decisions to be made
Layout and design	Easy-to-use guides (e.g., brochures)	Crisp, clear decisions (e.g., claims letters)

Writing Techniques for Audience Groups

Regardless of inherent limitations, categorizing the primary audience into one of these four basic groups—layperson, administrator, practitioner, or researcher—allows the writer to plan documents more effectively while prewriting. These audience categories suggest potential techniques to consider (Table 2.2), which will be examined more fully in subsequent chapters.

Laypersons not versed in health sciences, for example, generally prefer a more informal tone. Use of contractions and the second person (i.e., speaking to "you") can increase their comfort level, as can anecdotes or personal stories. Administrators need business-related facts quickly and concisely. Documents for administrators require unambiguous usage of terms such as "exclusions from coverage" and "indemnity" regarding an insurance claim or other business decision. Practitioners applying knowledge, as when diagnosing a condition, benefit from precise word choices, such as "drug-drug interactions" for an unexpectedly high laboratory result. Bulleted

AUDIENCE CATEGORY	
Practitioner	**Researcher**
Family physicians, nurses, technologists	Experts in university or institutional settings
Physician assistants Nurse practitioners Clinical pharmacists Allied-health workers	Research oncologists Nursing specialists Medicinal biochemists Molecular geneticists

AUDIENCE CATEGORY	
Practitioner	**Researcher**
Practical application, rather than theory	Theory and strategy, rather than application
Technical content directed to application	Complex arguments, statistics, references
Numbered steps or lists (e.g., manuals)	Scientific structure (e.g., journal articles)

lists that quickly provide the major decision points, along with checklists and flowcharts, facilitate their work. Finally, researchers expect literature references, complex graphics, and strong arguments so that they can assess the information. Their more critical approach often calls for complicated sentence structures that reflect nuances of the healthcare information.

Let's consider document examples for each audience type. To focus on the writing rather than the science, these four examples relate to one healthcare issue—management of risk factors for breast cancer.

Layperson-Focused Example

The excerpt in Figure 2.2 is from a layperson's version of information on breast-cancer screening from the website of the National Cancer Institute.[2] The sentence structure is clear and concise, especially when defining key terms: "A mammogram is an x-ray of the breast." A large schematic of a woman undergoing mammography captures attention while informing those unfamiliar with the procedure by showing equipment and even a real scan. Additionally, the schematic may put patients at ease by letting them know that the procedure is straightforward and non-invasive. Reflecting multicultural sensitivity—as well as the higher risk of breast cancer in certain demographic groups—this schematic portrays an African-American woman of an older age. This document is one of many related documents provided by the National Cancer Institute regarding breast cancer.

Figure 2.2 Layperson-Focused Example on Breast Cancer

Mammogram

Mammography is the most common screening test for **breast cancer**. A **mammogram** is an **x-ray** of the **breast**. This test may find tumors that are too small to feel. A mammogram may also find **ductal carcinoma in situ** (DCIS). In DCIS, there are **abnormal cells** in the lining of a breast **duct**, which may become invasive cancer in some women.

Mammograms are less likely to find breast tumors in women younger than 50 years than in older women. This may be because younger women have denser breast **tissue** that appears white on a mammogram. Because tumors also appear white on a mammogram, they can be harder to find when there is dense breast tissue.

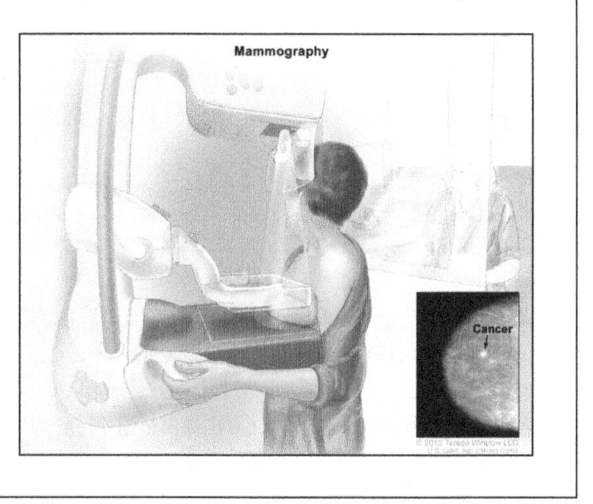

Administrator-Focused Example

The previous level of detail and overall style would not suit administrators, the next category of audience. A manager or claims processor for a healthcare insurance plan, for example, would need more specifics on the types, costs, and target groups for breast-cancer screening. An example is shown in the excerpt in Figure 2.3; these "story highlights" were a sidebar of information within a press release from the American Academy of Family Physicians, highlighting results from a recent study.[3] Formatting with bullets allows easy access to key points; all three support the administrator's need for information related to the costs of screening for breast cancer in older women. Given the study's finding of "insufficient evidence" in this age group, the third bullet concludes that "it remains unclear whether higher screening expenditures are achieving better breast cancer outcomes." An administrator needs to know that.

Figure 2.3 Administrator-Focused Example on Breast Cancer

STORY HIGHLIGHTS

- A study by researchers from the Yale University School of Medicine estimates that Medicare spends $410.6 million annually on breast cancer screening-related costs for women 75 and older, despite insufficient evidence to assess the benefits and harms of screening mammography in this age group.
- Regional variation in screening costs is substantial and often is driven by the use of newer and more expensive technologies.
- Although women residing in the "high screening-cost regions" were more likely than women in lower-cost regions to be diagnosed as having early-stage or in situ breast cancer, it remains unclear whether higher screening expenditures are achieving better breast cancer outcomes.

Practitioner-Focused Example

Guidelines (as in the adminstrator-focused excerpt) must be implemented if patients are to benefit. This is the role of practitioners, such as physicians and nurses. Recall, though, that practitioners have a high degree of knowledge in the subject area; the difference, relative to researchers, is the practitioner's focus on *application* of that knowledge. The excerpt in Figure 2.4, originating from the National Cancer Institute, is a version labeled for healthcare professionals.[4] Bolded headers and generous indentations facilitate quick access. Readers are assumed to recognize complex terms such as "comparator modality" and "population cohort studies"; however, "RCT" (randomized clinical trial) was defined earlier in the document. Referenced studies append details so that readers can judge for themselves the "Good" and "Poor" ratings of validity.

Figure 2.4 Practitioner-Focused Example on Breast Cancer

CLINICAL BREAST EXAMINATION

Benefits

Clinical breast examination (CBE) has not been tested independently; it was used in conjunction with mammography in one Canadian trial, and was the comparator modality versus mammography in another trial. Thus, it is not possible to assess the efficacy of CBE as a screening modality when it is used alone versus usual care (no screening activity).

Magnitude of Effect: The current evidence is insufficient to assess the additional benefits and harms of CBE. The single RCT comparing high-quality CBE to screening mammography showed equivalent benefit for both modalities. Accuracy in the community setting might be lower than in the RCT.

Study Design: Single RCT, population cohort studies.

Internal Validity: Good.

Consistency and External Validity: Poor.

Researcher-Focused Example

The fourth audience level—researcher—often necessitates the most complex of document approaches. The excerpt in Figure 2.5, the abstract of a full research article,[5] typifies the scientific detail demanded by healthcare researchers. Terms that might at first appear as jargon, such as "in situ hybridization" and "HER2/neu gene amplification" (among others) require no definitions. Regardless of this otherwise-obtuse style, the authors have energized the writing with the phrase "hijacks the skyrocketing cost" in the concluding sentence of the abstract.

Figure 2.5 Researcher-Focused Example on Breast Cancer

ABSTRACT

Few data on Human Epidermal Growth Factor Receptor 2 (HER2)-positive breast carcinomas have been reported for screen-detected breast carcinoma. Assessing the impact of a targeted intervention with anti-HER2 inhibitors on costs is required in order to plan for better strategies in screening programs. A total of 54,472 women were screened and 323 cases were found to be invasive cancer. We performed immunophenotypical-fluorescent in situ hybridization (FISH) analysis. Among 153 evaluable breast carcinomas, tumours displayed a 3+ scoring status in 16 (10%), 2+ in 12 (8%), 1+ in 29 (19%) and 0 in 96 (63%) of cases, respectively. All 3+ HER2+ cases and 2/12 2+ (17%) cases exhibited HER2/neu gene amplification, the remaining cases did not. In contrast to the higher incidence reported at the population level, 20-30% HER2-positive cases for metastatic carcinomas, and only 11% of the screen-detected breast carcinomas displayed HER2/neu gene amplification. Breast cancer detection by screening programs hijacks the skyrocketing cost of the use of targeted therapy in HER2-positive carcinoma.

KEYWORDS:
HER2, Screen-detected breast carcinoma, cost-effectiveness, gene amplification, screening, targeted therapy

Focus on Non-Researcher Audiences

Not only because of this complexity but also due to the historic emphasis on just this one group, writing for the researcher audience has already been the topic of numerous books. This book, therefore, focuses instead on the first three audiences in healthcare: laypersons, administrators, and practitioners. Each group is the subject of its own chapter later in the book (Chapters 9-11).

Venturing further into those territories, though, depends on understanding the building blocks used in preparing documents that target audiences for healthcare and other types of detailed information. The next several chapters explore those building blocks in more detail.

Chapter Summary

- Healthcare documents must be written with the primary audience (intended user) in mind; secondary audiences often require separate documents.
- Knowledge level for the healthcare topic reveals the techniques that may be most effective in that particular writing situation.
- Examination of many document examples helps writers to plan the writing of their own healthcare documents in the future.

Exercises for Practice

1. Identify the audience group for each of the following examples:

   - The **parents** of a child who has just been diagnosed with asthma
   - The **receptionist** at the laboratory that conducts allergen testing
   - The **physician's assistant** who administers the child's injections
   - The **scientist** at the pharmaceutical firm that devised the new inhaler
   - The **claims adjustor** who needs documentation for this therapy
   - The **allergist** who implements a treatment regimen for the child
   - The **child** who wonders why she runs out of breath when playing

2. Obesity is becoming a major health problem for the United States, as well as for other developed countries. Mounting evidence links obesity with higher risks of other diseases such as diabetes and hypertension. Moreover, obesity is not just a disease affecting adults—today, more and more children are classified as obese. Pretend that you work for a corporation that has recently opened an on-site health facility and offers reduced prices to encourage greater employee participation. Identify a document type (e.g., letter, booklet, or brochure) that you might need to write for each of the audience groups introduced in this chapter. How might you approach each document in terms of writing techniques according to the audience's knowledge level?

3. Select several examples of healthcare documents that you've received, such as a leaflet on a prescription drug, a list of approved providers from your insurance policy, or perhaps a story in your local newspaper. Into which audience group do you fall? How well do these documents work for you as the target audience?

References

1 Markel M. *Technical Communication*. 10th ed. New York: Bedford/St. Martin's; 2012.

2 National Cancer Institute. *Breast Cancer Screening (PDQ®), Patient Version*. Bethesda, MD: US National Institutes of Health; May 31, 2013. [Illustration © 2010 Winslow T; US Government has certain rights.] http://www.cancer.gov/cancertopics/types/breast.

3 American Academy of Family Practitioners. Breast Cancer Screening in Older Women Proving Costly, Likely Ineffective. *AAFP News Now*. Leawood, KS: American Academy of Family Physicians; 2013. http://www.aafp.org/news/health-of-the-public/20130116bcscreeningstudy.html.

4 National Cancer Institute. *Breast Cancer Screening (PDQ®), Health Professional Version*. Bethesda, MD: US National Institutes of Health; June 21, 2013. http://www.cancer.gov/cancertopics/types/breast.

5 Brunelli M, et al. HER2/neu gene determination in women screened for breast carcinoma: How screening programs reduce the skyrocketing cost of targeted therapy. *Anticancer Research* 2013; 33(9):3705-10. Available from: PubMed, Bethesda, MD.

3

SIMPLE STRATEGIES BY PURPOSE AND CONTEXT

Chapter Objectives

In this chapter, you will learn to:

- Interrelate audience, purpose, and context during prewriting
- Understand strategies used to organize healthcare documents
- Choose and apply simple strategies for uncomplicated topics

Audience, Purpose, and Context

Now that we're firmly in the prewriting stage, we've determined the target audience to be reached: layperson, administrator, or practitioner. (Later chapters expand on these groups by developing composites.) Even so, we're not ready to write yet. Although audience is a critical aspect of prewriting, we also need to determine *purpose*—what do we intend our document to do with regard to our audience? A concept related to purpose is *context*—in which situation(s) will our audience use this document?

Because healthcare documents are structured to achieve an end for the audience, a document's purpose is best expressed by the infinitive ("to") form of a verb to reflect that active goal. For example, a document could be meant simply *to describe* symptoms of a malady for a newly diagnosed patient. Another could be meant *to convince* a manager to switch insurance plans. A third might be intended *to teach* a technologist how to operate a new instrument. Note how the infinitive verbs succinctly express purpose.

Despite a potentially wide variety, purposes for healthcare writing generally fall into three broad categories:

- *To inform* (e.g., to define, to describe, to show)
- *To instruct* (e.g., to explain, to teach, to demonstrate)
- *To persuade* (e.g., to convince, to recommend, to change)

Table 3.1 lists various combinations of audience and purpose to exemplify how any one topic can be handled in several ways. The purpose of this list is to provide examples, not to delineate the only permutations possible; you'll continue to add to this list as you write more of your own healthcare documents.

Table 3.1 Examples of Healthcare Combinations by Audience and Purpose

TOPIC	AUDIENCE	PURPOSE
Healthcare Reform	Layperson	*To describe* the changes in legislation
	Administrator	*To instruct* how to process new forms
	Practitioner	*To convince* that changes are needed
Weight Control	Layperson	*To persuade* that daily exercising is good
	Administrator	*To define* therapies covered with plan
	Practitioner	*To explain* options of diet and exercise
Glucose Monitor	Layperson	*To demonstrate* use of new instrument
	Administrator	*To recommend* policy reimbursement
	Practitioner	*To show* changes from old to new model

One more prewriting condition needs to be addressed: context. Think of context as expressing the situation in which the audience will use the document. If someone with a severe allergy needs non-frequent but quickly discernible instructions for injecting epinephrine, then a laminated, wallet-sized card might be ideal. An insurance agent who processes reimbursement for hospital charges related to anaphylactic shock, in contrast, requires clear definition of the conditions that can lead to visiting the emergency room, as well as a breakdown of coverage levels for different plan options; a reference manual on the web might be most useful. Finally, a school nurse assisting a child recently stung by a bee needs a clear protocol to follow, supplemented by details related to potential liability for an unfortunate scenario in which the child is not treated promptly and properly. More than one document might be needed.

The one-size-fits-all approach will not routinely work, as shown in the previous example on treatment of allergic reactions. A person at risk might need something small and discrete for keeping close at hand, an insurance adjustor might prefer online details when processing claims in an office, and a school nurse might need one visual guide for the emergency and a second hardcopy document for indemnity. Context captures nuances that lead the writer to make the necessary decisions about how to proceed.

By considering audience in tandem with purpose and context, a writer is better prepared for the next step in the prewriting phase: strategy selection. A notable point about context is that, unlike audience and purpose, it may not be easily captured by one or two words. Use whatever format—from a few adjectives to a brief paragraph—that best captures the context for you.

Input Conditions into Output Parameters

So far, we've addressed the three input conditions—audience, purpose, context—that writers determine during the prewriting phase. Now comes the transformation that challenges the writer much like a jigsaw puzzle: how to assemble the document from the individual pieces in the box. Critical analysis leads the writer from input conditions to output parameters describing the final picture. Writers need to consider three key output parameters: medium, content, and strategy.

The purpose and context already identified tend to signal the *medium*, which is the form or genre of the document. Such genres can vary from a laminated card to a thick textbook to an online website. Audience, though, is important. Consider an at-risk family in an economically deprived section of town. Would they have consistent (or any) access to a computer? Might the text size on a printed brochure be too small for easy reading by a senior citizen? As discussed in later chapters, writers must consider an increasingly diverse populace, with due awareness of those for whom English is a second language. Literacy levels also come into play, whether for reading overall or for health-related comprehension in particular.

These types of consideration naturally lead to *content*—the amount and level of information to be provided. Audiences more knowledgeable about a healthcare topic may wish to have as much information available as possible; some may need that information provided in smaller chunks; others might be scared off by too much detail. Writers need find only the level and types of information (from the tremendous reams available) that will resonate with the intended audience. (Critically assessing that information is key, as we'll discuss in a subsequent chapter.)

Finally, we come to the last fundamental aspect in prewriting: *strategy*. Basically, strategy refers to deliberate choices made by writers in order to organize and present information in a particular pattern. Some of these patterns are simple and straightforward because the document is short and to the point. Other strategies are more complex if topics require an internal hierarchy with subsections; in fact, long documents often need nested strategies (i.e., separate strategies for each subsection). Those more complicated situations will be handled in the next chapter. For now, we'll focus on simple strategies that can be applied in more ways than their "simple" name might suggest.

Simple Strategies for Simple Situations

As just mentioned, we can often use simple organizational patterns—strategies—for simple situations. Why unnecessarily complicate healthcare documents? After all, we want our readers to focus on the content; our writing should be transparent in that it should lead the audience through information without distraction. Table 3.2 describes four common strategies for simple situations, along with examples of how they might be used. We'll explore each of these simple strategies now, using a fact sheet on the human circulatory system (Figure 3.1) as our example.[1]

Table 3.2 Descriptions and Examples of Simple Strategies for Healthcare Documents

SIMPLE STRATEGY	DESCRIPTION	EXAMPLE
Spatiality	Approaches by dimensions (e.g., bottom to top)	Structure of a human skeleton
Chronology	Orders a series either forward or reverse in time	Prenatal developmental stages
Specificity	Gives overview to familiarize before particulars	Treatment options in oncology
Importance	Lists issues from most to least relevant (or vice versa)	High cost of prescription drugs

Spatiality

A spatiality strategy describes in dimensional terms, since "spatiality" as a word refers to "space." Using a spatiality strategy, a writer emphasizes how a physical entity appears. A heart, for example, could be described by its four chambers. However, it could alternatively be described from the outside inward—first the outer pericardium or sac that envelops the heart; then the fist-like muscle itself; next the

Figure 3.1 Simple Strategies Exemplified by the Circulatory System

CIRCULATORY SYSTEM
All cells in the body need to have oxygen and nutrients, and they need their wastes removed. These are the main roles of the circulatory system. The heart, blood and blood vessels work together to service the cells of the body. Using the network of arteries, veins and capillaries, blood carries carbon dioxide to the lungs (for exhalation) and picks up oxygen. From the small intestine, the blood gathers food nutrients and delivers them to every cell.

Blood
Blood consists of:

Red blood cells - to carry oxygen
White blood cells - that make up part of the immune system
Platelets - needed for clotting
Plasma - blood cells, nutrients and wastes float in this liquid.

The heart
The heart pumps blood around the body. It sits inside the chest, in front of the lungs and slightly to the left side. The heart is actually a double pump made up of four chambers, with the flow of blood going in one direction due to the presence of the heart valves. The contractions of the chambers make the sound of heartbeats.

The right side of the heart
The right upper chamber (atrium) takes in deoxygenated blood that is loaded with carbon dioxide. The blood is squeezed down into the right lower chamber (ventricle) and taken by an artery to the lungs where the carbon dioxide is replaced with oxygen.

The left side of the heart
The oxygenated blood travels back to the heart, this time entering the left upper chamber (atrium). It is pumped into the left lower chamber (ventricle) and then into the aorta (an artery). The blood starts its journey around the body once more.

Blood vessels
Blood vessels have a range of different sizes and structures, depending on their role in the body.

Arteries
Oxygenated blood is pumped from the heart along arteries, which are muscular. Arteries divide like tree branches until they are slender. The largest artery is the aorta, which connects to the heart and picks up oxygenated blood from the left ventricle. The only artery that picks up deoxygenated blood is the pulmonary artery, which runs between the heart and lungs.

Capillaries
The arteries eventually divide down into the smallest blood vessel, the capillary. Capillaries are so small that blood cells can only move through them one at a time. Oxygen and food nutrients pass from these capillaries to the cells. Capillaries are also connected to veins, so wastes from the cells can be transferred to the blood.

Veins
Veins have one-way valves instead of muscles, to stop blood from running back the wrong way. Generally, veins carry deoxygenated blood from the body to the heart, where it can be sent to the lungs. The exception is the network of pulmonary veins, which take oxygenated blood from the lungs to the heart.

Blood pressure
Blood pressure refers to the amount of pressure inside the circulatory system as the blood is pumped around.

Common problems
Some common problems of the circulatory system include:

Aneurysm - a weak spot in the wall of an artery
Atherosclerosis - a narrowing of the arteries caused by plaque deposits
Heart disease - lack of blood supply to the heart because of narrowed arteries
High blood pressure - can be caused by obesity (among other things)
Varicose veins - problems with the valves that stop blood from running backwards.

[section deleted ...]

Things to remember

The circulatory system delivers oxygen and nutrients to cells and takes away wastes.
The heart pumps oxygenated and deoxygenated blood on different sides.
The types of blood vessels include arteries, capillaries and veins.

[section deleted ...]

Want to know more?
Go to More information for support groups, related links and references.

individual chambers; finally the walls separating the chambers and the valves connecting them.

In Figure 3.1, the short paragraph titled "The heart" applies a spatial description. The heart's position in the chest is shown through its orientation relative to other organs. Next revealed is the inner structure of this "double pump" that comprises four chambers whose contractions produce "the sound of heart beats." If appropriate for the audience, this spatial description could be further enhanced by describing color, texture, size, and other dimensions of the heart in even more detail; a graphic version of the heart would complement this text (see Chapter 7).

Worth noting is the variety of approaches that even a spatiality strategy affords. Imagine applying a spatiality strategy to describing a hospital: possibilities include outer shell to internal structures; bottom floor to top floor; admitting desk to each department or ward, to checkout desk; and so on. In other words, don't assume that a simple strategy limits your writing. Which of the various approaches works best for your audience and purpose? As the writer, that's your decision to make during prewriting—when you take the time to analyze your writing situation.

Chronology

Another typical strategy for simple situations is chronology: describing how a process occurs in time. How a new patient checks into a hospital fits into a chronology strategy. Other examples include procedures for laboratory tests on blood samples, pre-op checks before surgery, and quality-control points for the hospital's database. Note that these examples would be best described in *forward* chronological order, that is, from the first step to the last step.

In other instances, a variation using *reverse* chronological order might be a better choice. Suppose a patient develops a nosocomial infection (one picked up during a stay in the hospital). To discover when the infection might have originated, the nursing staff may begin at the first sign of the patient's infection and then work backward in time toward a likely point of exposure. A training manual for investigating nosocomial infections thus might be better constructed in a reverse chronology strategy. Once again, the choice is yours, depending on the audience and purpose.

In the example of Figure 3.1, a chronology strategy appears in the two sections on the heart's right and left sides. Although the four chambers could have been described spatially, the writer opted to track the flow of blood through the heart. Since the heart's chambers have already been described spatially, the chronological approach builds on the audience's spatial perspective of the heart to illustrate blood flow, including the related process of gas exchange within the lungs. The concluding sentence indicates the cyclical nature of the flow of blood: "The blood starts its journey around the body once more."

Specificity

This last statement quoted above from Figure 3.1 transitions easily into subsequent descriptions of blood vessels, including arteries, capillaries, and veins. These descriptions begin with an overview of the sizes and structures of various blood vessels; in other words, the writer provides a general overview before the subsequent descriptions of specific types of blood vessels. Such an approach captures the essential aim of the specificity strategy: providing an overview that orients the audience before exposing them to the specific elements that follow.

In the example text, note how the writer intentionally builds on the general theme of blood flow to order the three classes of blood vessels. Leading directly from the heart, arteries carry oxygenated blood (except for the pulmonary artery); then the larger arteries subdivide into smaller branches like a tree—illustrating a simile as another writing tool. (*Similes* compare through statements beginning with "like" or "as"; the related *metaphors* more directly link the two items, as in calling a heart the combustion engine of the body.) Capillaries, the smallest group of vessels, follow because they connect arteries to the final group, the veins.

In most instances, a specificity strategy provides an overview before the details. However, at times, a writer may wish to give details before giving the generalities, much like challenging an audience to solve a puzzle from individual clues. Although rare, this reverse specificity relates more to inductive reasoning (covered in Chapter 6).

Importance

The last simple strategy is importance: ordering units, ideas, or elements by their relevance to the audience. In Figure 3.1, an importance strategy organizes bulleted items under "Things to remember." The most important takeaway message for the intended audience is the function of the circulatory system. The subsequent two points make sense only if this key idea is first grasped and retained. Of course, three items do not provide too many opportunities for organization. Consider, though, a more detailed overview of the heart, this time targeting medical students; certainly, a longer list of ideas to remember would require a logical organizational plan to facilitate this audience's comprehension of such complex material.

Note, however, that not all lists require an importance strategy. The section titled "Blood" lists blood's four components—red cells, white cells, platelets, and plasma—in no clear order (other than perhaps familiarity to the reader). However, since the purpose of this fact sheet is to discuss the circulatory system, particularly its function in the body, then beginning with red cells would be most important (but plasma should then follow). Another list occurs in the section termed "Common problems";

here, the writer chose an alphabetical listing of problems, perhaps to allow subsequent access in an index format. Regardless, starting with "heart disease" as a general concept before adding any specific conditions might be more useful for a layperson audience.

Overall, then, this one brief document exemplifies effective use of four different strategies for simple situations: spatiality, chronology, specificity, and importance. As shown, these basic, simple strategies can be varied to satisfy the needs of audience and purpose. The use of the adjective "simple" belies the various manners in which astute writers can apply these strategies according to the input conditions examined during the prewriting phase. The possibilities broaden for documents tackling complex information, as discussed in the next chapter.

Chapter Summary

- During prewriting, writers need to identify not only the target audience but also the document's intended purpose and the context in which it will be used.
- Four simple strategies—spatiality, chronology, specificity, importance—provide various ways to present readers with fairly straightforward information.
- Combinations of these simple strategies can be used to organize smaller but important details within an overall document of relatively short length.

Exercises for Practice

Identify the input conditions and output parameters for the following examples from healthcare documents. Link audience, purpose, and context (input) strategies used in the excerpts. Justify your assessments.

1. Asthma is a chronic condition whose long-term management includes drugs from different pharmaceutical classes. Selection of appropriate medications depends on the patient's status, as detailed in guidelines from an expert panel of researchers; the excerpt in Figure 3.2 comes from a section on pharmacological therapy.[2]

Figure 3.2 Long-Term Management of Asthma

Managing Asthma Long Term

Achieving and maintaining asthma control requires four components of care: assessment and monitoring, education for a partnership in care, control of environmental factors and comorbid conditions that affect asthma, and medications. A stepwise approach to asthma management incorporates these four components, emphasizing that pharmacologic therapy is initiated based on asthma severity and adjusted (stepped up or down) based on the level of asthma control. Special considerations of therapeutic options within the stepwise approach may be necessary for situations such as exercise-induced bronchospasm (EIB), surgery, and pregnancy.

2. Educational requirements dictate much of a student's academic day; sometimes, though, critical topics like health education may not be sufficiently emphasized. The excerpt in Figure 3.3 comes from a government website that provides schools with important information related to the teaching of health.[3]

Figure 3.3 Introductory Text on Physical Fitness

Health Education Curriculum Analysis Tool (HECAT)

The Health Education Curriculum Analysis Tool (HECAT) can help school districts, schools, and others conduct a clear, complete, and consistent analysis of health education curricula based on the National Health Education Standards and CDC's Characteristics of an Effective Health Education Curriculum.

Results of the HECAT can help schools select or develop appropriate and effective health education curricula and improve the delivery of health education. The HECAT can be customized to meet local community needs and conform to the curriculum requirements of the state or school district.

The HECAT features:

- Guidance on using the HECAT to review curricula and using the HECAT results to make health education curriculum decisions
- Templates for recording important descriptive curriculum information for use in the curriculum review process
- Preliminary curriculum considerations, such as accuracy, acceptability, feasibility, and affordability analyses
- Curriculum fundamentals, such as teacher materials, instructional design, and instructional strategies and materials analyses
- Specific health-topic concept and skills analyses
- Customizable templates for state or local use
- Summary score forms for consolidating scores from the review of a single curriculum and for comparing scores across multiple curricula

3. Suicide has become a critical problem—and one not always openly discussed. Yet, through dissemination of facts on mental illness, persons at risk for suicide may be recognized and helped. The excerpt in Figure 3.4 from a government webpage lists protective factors that may help at-risk individuals.[4]

Figure 3.4 Factors Protecting against Suicide

Protective Factors for Suicide

Protective factors buffer individuals from suicidal thoughts and behavior. To date, protective factors have not been studied as extensively or rigorously as risk factors. Identifying and understanding protective factors are, however, equally as important as researching risk factors.

Protective Factors

- Effective clinical care for mental, physical, and substance abuse disorders
- Easy access to a variety of clinical interventions and support for help seeking
- Family and community support (connectedness)
- Support from ongoing medical and mental health care relationships
- Skills in problem solving, conflict resolution, and nonviolent ways of handling disputes
- Cultural and religious beliefs that discourage suicide and support instincts for self-preservation

References

1 Better Health Channel, State of Victoria. *Circulatory System* [fact sheet]. Victoria, Australia: Victorian Ministry of Health; 2013. http://www.betterhealth.vic.gov.au/bhcv2/bhcarticles.nsf/pages/Circulatory_System.

2 National Heart, Lung, and Blood Institute. *Expert Panel Report 3 (EPR-3): Guidelines for the Diagnosis and Management of Asthma—Summary Report 2007.* Bethesda, MD: US Department of Health and Human Services; October 2007:15. NIH Publication No. 08-5846. http://www.nhlbi.nih.gov/guidelines/asthma/asthsumm.htm.

3 Centers for Disease Control and Prevention, US Department of Health and Human Services. *Health Education Curriculum Analysis Tool, 2012.* Atlanta: CDC; 2012. http://www.cdc.gov/HealthyYouth/HECAT/index.htm.

4 National Center for Injury Prevention and Control, Centers for Disease Control and Prevention. *Suicide: Risk and Protective Factors.* Atlanta: CDC; last updated September 18, 2012. http://cdc.gov/violenceprevention/suicide/riskprotectivefactors.html.

4

COMPLEX AND NESTED STRATEGIES

Chapter Objectives

In this chapter, you will learn to:

- Identify complex strategies for organizing complicated topics
- Discern topics that benefit from a nesting of several strategies
- Apply signposting to orient readers to a document's structure

Complex Strategies for Complex Situations

In many instances, the increasing complexity of healthcare surpasses the capabilities of simple strategies. Consider the earlier overview of the circulatory system (Figure 3.1 in Chapter 3): because of the layperson audience, the writer intentionally kept the document at a simpler level. Certainly, a compilation of insurance plans for different cardiovascular conditions, needed by an administrator, would be more detailed. A textbook for medical students, moreover, would require excruciating attention to every nuance and detail that these nascent practitioners and researchers must master.

The following sections review three of the commonly used complex strategies. Table 4.1 describes these complex strategies: problem, method, solution; classification and partition; and comparison and contrast. Although any of these three can be applied alone within shorter documents, combinations of these complex strategies allow writers to organize longer documents, such as technical reports. For examples in this chapter, selections are provided from a multipage report from the US Centers for Disease Control and Prevention (CDC) exploring the threat of emerging resistance to antibiotics.[1]

Table 4.1 Descriptions and Examples of Complex Strategies for Healthcare Documents

STRATEGY	DESCRIPTION	EXAMPLE
Problem, method, solution	Clearly identifies problem, details steps for its investigation, and offers options to solve it	Public-service campaigns for increasing immunization rates
Classification and partition	Sorts many items into limited groups (classify) or divides complex whole into parts (partition)	Diseases by genetic factors Medicine into specializations
Comparison and contrast	Juxtaposes two or more wholes according to several characteristics of relevance	Predisposition of mental illness among different ethnic groups

Problem, Method, Solution

Common in publications for researcher audiences is the first complex strategy—*problem-method-solution.* This strategy mirrors the "IMRAD" (**I**ntroduction, **M**ethods, **R**esults, **A**nd **D**iscussion) structure typically used for research articles. Publications for researchers almost always include background on the topic (Introduction), followed by techniques to investigate the topic (Methods), and then findings from the investigation (Results and Discussion).

Nonetheless, this same complex strategy also can be used in documents for other healthcare audiences. The problem-method-solution strategy mimics that publication structure, but with three (rather than four) sections: an overview and rationale to set the issue into perspective, the ways in which the issue was examined, and recommendations for addressing the issue.

Problem-method-solution could easily apply to an analysis of the increasing costs of healthcare. For layperson, administrator, or practitioner audiences, this analysis would first put the problem into perspective (current healthcare shortcomings), next detail the method for analyzing plans (statistical cost projections), and then recommend one of the plans as a solution to this conundrum (government subsidies, for instance). Audiences are led through the same types of information but at a less challenging pace.

Overall, the CDC report utilizes the problem-method-solution strategy throughout the entire report. First, the global threat of emerging antibiotic resistance is established (problem); the Introduction provides this evidence, as shown in this excerpt:

> Antibiotic resistance is a worldwide problem. New forms of antibiotic resistance can cross international boundaries and spread between continents with ease. Many forms of resistance spread with remarkable speed. World health leaders have described antibiotic-resistant microorganisms as "nightmare bacteria" that "pose a catastrophic threat" to people in every country in the world.[1(p11)]

Beyond this introductory paragraph, the case for addressing this global threat continues with statistical and monetary data as evidence. In fact, establishment of the "problem" continues for seven pages of text and tables.

After thoroughly establishing the problem, the report transitions into the methods used for analyzing the detailed trends in emerging resistance; this analysis encompasses various microorganisms along with antibiotics available to combat them. Recapitulation of these complicated methods is beyond the scope of this chapter, but a sense is achieved through the text excerpt that follows:

> This report uses several methods, described in the technical appendix, to estimate the number of cases of disease caused by antibiotic-resistant bacteria and fungi and the number of deaths resulting from those cases of disease. The data presented in this report are approximations, and totals, as provided in the national summary tables, can provide only a rough estimate of the true burden of illness.... The actual number of infections and the actual number of deaths, therefore, are certainly higher than the numbers provided in this report.[1(pp18-19)]

The 14-page technical appendix details and justifies specific methods used for assessing the microorganisms included in this report.

As might be expected, solutions for addressing this global threat are multifaceted. These solutions comprise 16 pages (whose strategy is discussed below). Four key sets of solutions are identified, followed by elaboration of the manner in which these solutions should be implemented. The bulk of this report, in fact, consists of a long series of action plans for each of the microorganisms. However, the report team encapsulates the solution to these global threats in one succinct statement: "Bacteria will inevitably find ways of resisting the antibiotics we develop, which is why aggressive action is needed now to keep new resistance from developing and to prevent the resistance that already exists from spreading."[1(p12)]

Classification and Partition

The next complex strategy—classification and partition—represents two facets of a dual approach. Classification and partition apply when the topic involves many separate but related items requiring organization (classification) or a complicated conglomeration that must be broken down into manageable units (partition). Healthcare offers many opportunities for applying this strategy pair.

The key to using classification is recognizing that a large number of items must first be sorted before being described or analyzed. A typical situation for classification would be categorizing therapeutic agents such as prescription drugs: these categories could be by indication (e.g., asthma, hypertension, diabetes), mode of action (e.g., hormonal, antibiotic), or even payment (e.g., covered or excluded by a drug formulary). Depending on the audience and purpose, diseases could be classified by body system, hospitals by the services provided, and insurance plans by coverage options.

Partition comes from the opposite vantage: a complicated situation that cannot be tackled until broken into smaller, more manageable units. Healthcare examples include the medical delivery system, role of childhood immunization, and the nutrition pyramid. In the first case, the medical delivery system could break into primary care, emergency care, hospitalization, urban clinics, and rural delivery; however, if the audience's needs were different, the system could be divided by the type of provider (e.g., physician, nurse, pharmacist, and so on).

Once again, the CDC report provides real examples. Classification could certainly apply in organizing the multitude of microorganisms with potential antibiotic resistance. In this instance, the report team chose to focus on the severity of the threat, using groups of "urgent, serious, or concerning."[1(p20)]

Understandably, the recommended action plans are not only comprehensive but also multifaceted (as noted earlier). Hence, the report team employed the strategy of partition to present the solution as broken into four approaches: "1) preventing infections from occurring and preventing resistant bacteria from spreading, 2) tracking resistant bacteria, 3) improving the use of antibiotics, and 4) promoting the development of new antibiotics and new diagnostic tests for resistant bacteria."[1(p7)] The report's long section on these four approaches fleshes out the details.

Comparison and Contrast

The last set of our three complex strategies—comparison and contrast—should be familiar to anyone who has ever written an academic essay. Using this strategy, a writer discusses how two or more items are similar to, as well as different from, the others. The "similar to" piece refers to the comparative aspect of this strategy; "different from" refers to the contrast aspect. For healthcare topics, example uses of this complex strategy would include treatments for a disease, options for insurance coverage, parts of the human body, and many other possibilities.

In comparison and contrast, individual items being examined (such as spleen and bone marrow, both of which produce blood cells) are termed "wholes"; "parts" are then determined from criteria to be used in judging these items. For example, a discussion of spleen vs. bone marrow might consider location in the body, types of blood cells, and role in fighting infection. Often, wholes are easier to identify than parts: an employee clearly needs to compare and contrast four offered insurance plans, but on what characteristics other than monthly cost? Those characteristics—including access to care, need for prior authorization, and availability of providers—should be easily identifiable in insurance information provided by an employer.

Identifying the wholes and parts is still not enough; the order in which these items and criteria are addressed is critical. In other words, how do we order them into an outline for our document? This framework takes either of two forms:

- *Whole-by-whole* using the items to be examined as the major grouping
- *Part-by-part* using specific criteria for judgment as the major grouping

In the spleen vs. bone marrow scenario, a whole-by-whole framework would first discuss the spleen for each of the criteria, followed by a parallel discussion of the bone marrow. The advantage occurs when the overall physiological roles of spleen and bone marrow are needed, as for an audience unfamiliar with either. However, for an audience familiar with both organs, a head-to-head line-up by criteria would be able to differentiate their subtleties. While determining this framework, the writer concurrently generates an outline of the document and its sections.

If we return to the CDC report, we see that a whole-by-whole approach is used to assess the threats posed by microorganisms in terms of antibiotic resistance. The three classifications for the level of threat—urgent, serious, or concerning—represent the wholes; the parts are the "seven factors associated with resistant infections: clinical impact, economic impact, incidence, 10-year projection of incidence, transmissibility, availability of effective antibiotics, [and] barriers to prevention."[1(p20)] These parts then allow the wholes to be compared and contrasted meaningfully in terms of understanding and combatting the global threats posed by emerging antibiotic resistance.

Part-by-part approaches tend not to be used as often, though, because typically readers need to understand the units of discussion through whole-by-whole approaches. Nevertheless, part-by-part lends itself to particular situations in which readers instead focus on the criteria. Regarding the CDC report, each of the many antibiotics itself may be associated with adverse events, more commonly known as side effects. Possibilities of such adverse events, which could be discussed antibiotic by antibiotic (whole-by-whole) might be more meaningful if discussed by adverse event types (part-by-part). Here are those parts with relevance to the antibiotics under consideration:

- Allergic reactions—"These reactions can range from mild rashes and itching to serious blistering skin reactions, swelling of the face and throat, and breathing problems."
- Additional infections—"When a person takes antibiotics, good bacteria that protect against infection are destroyed for several months."
- Drug interactions—"Antibiotics can interact with other drugs patients take, making those drugs or the antibiotics less effective."[1(p26)]

Nested Strategies and Signposting

Using complex strategies, as just explored for the CDC report on the threats of emerging antibiotic resistance, may seem daunting. They require time and effort from a writer while still in the prewriting phase—even a writer eager to commence writing! As the discussion of the various sections of the CDC report highlights, however, healthcare topics can be complicated for audiences who are not experts in those fields. (These topics can be complicated for researchers, too.) Nonetheless, this initial investment while prewriting yields considerable payoffs in the writing and postwriting phases.

Indeed, as documents increase in organizational complexity, writers often employ *nested* strategies. Information on emerging antibiotic resistance, for instance, comprises a surfeit that could overwhelm a reader. Therefore, the writing team chose to *partition* the document into the separate sections that we've just analyzed. Then, each of these sections could be tackled with its own strategy; in fact, a large section may itself be subdivided into subsections, each with its own strategy (as just shown in this chapter's examples). In other words, each subsection requires its own strategy *nested* within an overall strategy for the full document. The context for each section is also critical, as different situations benefit from different approaches.

This nesting of strategies, typically used for different sections of a much longer document, gives the writer flexibility in achieving the purpose for the target audience. In complicated situations, the writer may need to apply more complex strategies. The astute writer, nonetheless, will not fully discount simple strategies. Often, complex strategies provide the large-scale framework for a document, whereas simple strategies tailor the subsidiary parts within that framework.

Another writing technique that complements nested strategies is *signposting*. Basically, a signpost is a group of words that alerts readers to a section's organization. Through signposting—which can be applied for any writing strategy—the writer guides the audience when the information might start becoming more complex and thus require more focus on the reader's part. Signposts within documents function like signposts on a road that alert drivers to slippery roads, falling rocks, or speed bumps.

The CDC report on emerging resistance provides several examples of signposting. Sometimes, signposting can be overt, as in the subheading "Timeline of Key Antibiotic Resistance Events" preceding a graphic in chronological order.[1(p28)] In other instances, signposting statements can imply a strategy, as in the following sentence that introduces a large section whose information is grouped by partition: "This section focuses on CDC's works to prevent antibiotic-resistant infections in healthcare settings, in the community, and in food."[1(p32)] A strategy specific to the topic, moreover, often needs identification: "The table below describes the drug classes used to treat these infections.... The classes are in order of most likely to be used to less likely to be used."[1(p22)] Techniques like these gently guide readers through the complicated report.

As demonstrated throughout this chapter and the previous chapter, selection of strategies during the prewriting phase—particularly for longer documents—identifies a framework on which to organize information. The benefits of signposting accrue not just to the reader but also to the writer: choosing strategies before the actual writing begins sets up an outline. In this outline, headings and subheadings reflect those strategies. As the document unfolds, of course, writers should allow themselves the freedom to modify such outlines as appropriate.

When writers identify strategies during prewriting, the document begins to write itself. Before you realize it, you'll find yourself jumping hurdles of writer's block that sometimes stymie even experienced writers. With the outline structured, you'll be ready to collect information to fill those sections. Finding sources for documents is the focus of the next chapter.

Chapter Summary

- As topics increase in complexity, writers should consider organizing information through one of several complex strategies: problem, method, solution; classification and partition; and comparison and contrast.
- Strategies identified during the prewriting phase allow the writer to outline the document into pieces that meet the needs of the intended audience. Purpose and context also come into play.
- Larger documents often benefit from using nested strategies: each section uses its own appropriate strategy.
- Writers can orient readers to the text's strategies through statements that signpost the information to follow. Readers then know what to expect.

Exercises for Practice

Identify the strategies used in the following examples from healthcare documents. Justify your answers by highlighting key terms, organizational signposting, or other techniques from this chapter. Note that these excerpts are from longer documents; checking the full text may provide clues on nested strategies (see reference list for the hyperlinks). Identify the audience and purpose for each document.

1. Despite spending the highest per capita on healthcare, the US does not achieve equitable health across its various subgroups. Notably, racial and ethnic minorities often face higher risks of particular diseases, as highlighted in this government excerpt (Figure 4.1) from the REACH program.[2]

Figure 4.1 Racial and Ethnic Disparities in Health

Why is eliminating health disparities important?

The CDC believes every person should have the opportunity to attain his or her full health potential. CDC seeks to eliminate barriers to achieving this potential because of social position or other socially determined circumstances. Health disparities remain widespread among members of racial and ethnic minority populations.

- Heart disease is the leading cause of death for people of most ethnicities in the United States.
- Non-Hispanic blacks have the highest rates of obesity (44.1%) followed by Mexican Americans (39.3%).
- Compared to non-Hispanic whites, the risk of diagnosed diabetes is 18% higher among Asian Americans, 66% higher among Hispanics/Latinos, and 77% higher among non-Hispanic blacks.

2. As evidenced by news reports, the incidence of overweight persons in the US, as well as other Western countries, has climbed to alarming rates. One way to tackle this issue is to educate school-age children on the benefits of consuming fruits and vegetables. The excerpt in Figure 4.2 discusses such school programs.[3]

Figure 4.2 Promotion of Healthy Diets through School-Based Programs

Strategy 9. Establish policies to incorporate fruit and vegetable activities into schools as a way to increase consumption

DEFINITION

To reinforce health messages, schools can establish policies to incorporate activities that involve fruits and vegetables into their curricula. Such activities include gardening, agricultural education (e.g., visits to farms), lessons on fruit and vegetable preparation, and tasting demonstrations. School policies also can encourage integrated approaches to these activities, where the produce from school gardens is used as part of classroom activities, as well as in food service venues and at events and fundraisers held at the school.

These curriculum-based activities provide students with hands-on experiences with fruits and vegetables and support policy and environmental changes in the school setting. These activities also can be part of a farm-to-school program, or they can be designed to encourage the school to adopt such a program. Examples of policy and environmental changes include applying for the USDA's Free Fruit and Vegetable Snack Program, signing up for the *Let's Move! Salad Bars to Schools* program to help add a salad bar to the school cafeteria, or creating standards for competitive foods that require fruit and vegetable options.

In school gardening programs, students participate in growing and harvesting a variety of fruits and vegetables. Gardening activities provide hands-on study of nutrition and science concepts, as well as ecology, math, history, social science, and the visual arts.

Agricultural education provides students a chance to learn about fruit and vegetable production at school, in the community, and elsewhere. Classroom activities that involve fruits and vegetables can teach students how to select and prepare fruits and vegetables and encourage them to taste and handle both familiar and unfamiliar varieties. Food preparation classes can support experiential learning about nutrition, health, science, and ecology.

3. Citizens expect their government agencies to provide protection against risk to public health, such as the spread of foodborne illnesses. The excerpt in Figure 4.3 provides definitions of different categories of foodborne outbreaks.[4] In addition to identifying input conditions and output parameters, outline a larger manual based on this information. Include nested strategies and draft signposting statements.

Figure 4.3 Comparison of Foodborne Outbreaks

Size and Extent of Foodborne Outbreaks

When two or more people get the same illness from the same contaminated food or drink, the event is called a foodborne outbreak. Illnesses that are not part of outbreaks are called "sporadic." Public health officials investigate outbreaks to control them, so more people do not get sick in the outbreak, and to learn how to prevent similar outbreaks from happening in the future.

The size and scope of a foodborne outbreak can vary based on which pathogen or toxin is involved, how much food is contaminated, where in the food production chain contamination occurs, where the food is served, and how many people eat it. For example:

- **Small, local outbreak**—A contaminated casserole served at a church supper may cause a small outbreak among church members who know each other.
- **Statewide or regional outbreak**—A contaminated batch of ground beef sold at several locations of a grocery store chain may lead to illnesses in several counties or even in neighboring states.
- **Nationwide outbreak**—Contaminated produce from one farm may be shipped to grocery stores nationwide and make hundreds of people sick in many states.

References

1 Centers for Disease Control and Prevention. *Antibiotic Resistance Threats in the United States, 2013.* Atlanta: US Department of Health and Human Services; 2013; last updated March 10, 2014. http://www.cdc.gov/drugresistance/threat-report-2013/index.html.

2 Division of Community Health, National Center for Chronic Disease Prevention and Health Promotion. *Racial and Ethnic Approaches to Community Health (REACH): Investments in Community Health.* Atlanta: Centers for Disease Control and Prevention; last updated October 19, 2012. http://www.cdc.gov/reach.

3 Centers for Disease Control and Prevention. *Strategies to Prevent Obesity and Other Chronic Diseases: The CDC Guide to Strategies to Increase the Consumption of Fruits and Vegetables.* Atlanta: US Department of Health and Human Services; 2011. http://www.cdc.gov/obesity/downloads/FandV_2011_WEB_TAG508.pdf.

4 Centers for Disease Control and Prevention. *Size and Extent of Foodborne Outbreaks.* Atlanta: US Department of Health and Human Services; last updated September 27, 2011. http://www.cdc.gov/foodsafety/outbreaks/investigating-outbreaks/size-extent.html.

5

RELIABLE SOURCES OF INFORMATION

Chapter Objectives

In this chapter, you will learn to:

- Access and assess sources of healthcare information
- Identify traditional vs. electronic information sources
- Conduct searches for information more expeditiously
- Credit quoted and paraphrased sources by referencing

Accessing and Assessing Healthcare Sources

We've identified the target audience, we've determined the document's purpose, and we've outlined our document through overall and nested strategies. The document remains in the prewriting phase, though, until we've found reliable sources of healthcare information. The information gleaned from those sources will populate the outline when the document moves into the writing phase.

In many ways, today's expansive availability of reference sources—especially the omnipresent Internet—makes this aspect of healthcare writing much easier. With clicks and scrolls, writers can take advantage of a wide array of sources that local libraries may not even have in hard copy. Information is also updated (often instantaneously) so that writers always have the most current healthcare findings at hand. True, some specialized Internet sites require an access fee, but those fees are typically offset through incredible savings (of both money and time) from interlibrary loans and personal travel.

This incredible availability of information, nonetheless, comes at another price: now that writers are *accessing* healthcare (and virtually all) information with minimal effort, writers must expend more effort on *assessing* those sources for reliability. Writers often learn that trade-off the hard way. A quick search through any of the more common search engines (e.g., Bing, Google, Yahoo, to name just a few) can provide countless hits leading to information with some relevance to the topic, but many (if not most) of these hits may not provide the level of information needed for the document's audience and purpose. Moreover, too many hits that may appear worthwhile require deeper scrutiny into their overall reliability.

In other words, writers cannot be satisfied with increased accessibility because anyone with the know-how can upload information without any guarantee of its validity. Even excluding the sad truth that some individuals would knowingly upload information to harm others, the real possibility exists that others might unknowingly post outdated or simply inaccurate information. In the Internet Age, writers need to shift their focus from accessing to assessing healthcare information.

Along with this evolution of the responsibilities for writers comes an evolution of the roles of librarians. The following advice for evaluating online sources originated with reference librarians at a university.[1] These four criteria for evaluating online sources can be adapted for all sources, including those in hard copy:

- *Authority* Can the author be trusted to have true expertise in this field?
- *Accuracy* Does the information derive from verifiable sources?
- *Currency* Are relevant dates identified for creation and update?
- *Objectivity* How does the source maintain fairness without bias?

Authority refers to the qualifications of those who produced the information or sources. Do their credentials (which must always be available) document their expertise for the particular topic? Moreover, when an organization is the author, would we worry that their view might be slanted toward the organization's goals? Links to a sponsoring organization should always be evaluated. Any organization wants to portray its case in as strong a manner as possible, but some claims can be overstated.

Accuracy is paramount. Although writers may not be as familiar with their topics as are researchers, writers are responsible for becoming conversant with their topics, particularly in a field like healthcare where inaccuracy can lead to harm. Even while climbing the learning curve, writers can look for clues of accuracy, such as reliable references. Warning flags can arise from the writing: errors in grammar and punctuation may reflect a lack of care with facts and data.

Currency, at first glance, might seem more of an issue for hard-copy sources that can quickly become outdated, but currency applies to online sources, too. All references should clearly date the information, whether for its original creation or its most recent update. Although online sources are assumed to be up-to-date, they can quickly become stale in an ever-changing field like healthcare. For online sources, another warning sign would be if links were no longer active.

Objectivity follows from authority. An author with stellar credentials may still be linked with an organization with a vested interest in the outcome. Many such biases are financial, such as promotional support from a drug manufacturer. Especially suspect are websites or journals peppered with advertisements, even though some may simply offset the cost of publication and maintenance.

In short, any source—hard-copy or online—must be assessed by the writer before being used in a healthcare document. Accessibility is simply not enough. Writers have an ethical responsibility to do all that they can to safeguard their audiences.

Traditional Sources in Hard-Copy Format

Despite the proliferation of online sources for all kinds of information—including healthcare—many helpful, if not critical, sources can be accessed in traditional hard-copy format. In fact, established writers should maintain their own small hard-copy library on a nearby bookshelf for easy access. That quick flick through paper can sometimes be faster and more rewarding than logging onto a new or even trusted site. Recommended types of references for the beginning writer's bookshelf fall into three types:

- Medical dictionaries
- Healthcare references
- Editorial style guides

Certainly, each writer needs to select the most appropriate versions from those available. Table 5.1 offers representative selections (not intended to be endorsements, but rather examples for consideration). Take time to peruse various options before investing money, as many of these specialized books can be expensive to buy, only to be updated within a few years. Some books, interestingly, are published in different versions so as to target physicians, nurses, or other groups, respectively. Knowing target audiences for the expected workload allows writers to spend their funds most parsimoniously—especially since funds for online services may also be needed.

Table 5.1 Representative Sources in Hard-Copy Format

SOURCE CLASS	REPRESENTATIVE EXAMPLES
Medical dictionaries	*Stedman's Medical Dictionary for the Health Professions and Nursing* defines over 56,000 medical terms, with online subscriptions available[1] *Taber's Cyclopedic Medical Dictionary* defines over 50,000 terms, many with illustrations, with online and mobile versions available[2]
Healthcare references	*The Merck Manual of Diagnosis and Therapy* covers a variety of diseases, disorders, injuries, symptoms, and therapies at a technical level[3] *Goodman & Gilman's The Pharmacological Basis of Therapeutics* covers medical pharmacology, including drug actions and drug-drug interactions[4] *Physicians' Desk Reference (PDR)* is a compendium of approved labeling for drugs and prescription information from pharmaceutical manufacturers[5]
Editorial style guides	*American Medical Association Manual of Style: A Guide for Authors and Editors* details publication standards and lists available resources[6]

1 Stedman TL. *Stedman's Medical Dictionary for the Health Professions and Nursing*. 7th ed. Philadelphia: Lippincott Williams & Wilkins; 2011.
2 Venes D. *Taber's Cyclopedic Medical Dictionary*. 22nd ed. Philadelphia: F.A. Davis Company; 2013.
3 Porter RS (ed.). *The Merck Manual of Diagnosis and Therapy*. 19th ed. Whitehouse Station, NJ: Merck Sharpe & Dohme Corp.; 2011.
4 Brunton L, Chabner B, Knollman B. *Goodman & Gilman's The Pharmacological Basis of Therapeutics*. 12th ed. New York: McGraw-Hill Professional; 2010.
5 PDR Staff. *Physicians' Desk Reference* (PDR). 67th ed. Montvale, NJ: PDR Network; 2012.
6 American Medical Association. *American Medical Association Manual of Style: A Guide for Authors and Editors*. 10th ed. New York: Oxford University Press, USA; 2007.

Expansive Sources in Electronic Format

As just discussed, the traditional sources in hard-copy format build a basis for the healthcare writer's bookshelf. That traditional bookshelf, however, must be supplemented with access to the expansive healthcare sources available online. These sources come in many varieties, with three most pertinent to the beginning writer:

- Commercially funded websites
- Research institution websites
- Government agency websites

Table 5.2 provides a sense (but again not an endorsement) of the many online sources accessible through the Internet. Although all online sites must be evaluated carefully for criteria presented at the start of this chapter, these tried-and-tested sites offer a baseline degree of reliability for the beginning writer.

Table 5.2 Representative Sources in Electronic Format

WEBSITE LINK	SITE DESCRIPTION
www.webmd.com	WebMD covers news, diseases, services, and guides on healthcare for general audiences, but with advertisements from healthcare firms
www.apha.org	Public Health Links, a collaboration between the American Public Health Association and agencies and libraries, provides resources for the public health workforce
www.medscape.com	Medscape, part of WebMD Health Professional Network, targets clinicians and other practitioners in facilitating diagnosis and improving patient care
www.hhs.gov	US Department of Health and Human Services provides information on diseases, emergencies, regulations, grants, aging, families, and more
www.fda.gov	US Food and Drug Administration provides information on foods and drugs (as well as biologics and cosmetics), grouped for patients, practitioners, scientists, and industry
www.cdc.gov/nchs	National Center for Health Statistics, provided by Centers for Disease Control and Prevention, links to quantitative data concerning various healthcare topics

Expeditious Searches for Information

With accessibility to such expansive sources, beginning healthcare writers must learn to conduct searches as expeditiously as possible; writers must know not only the types of sources needed, but also the best means to find those sources. Any search relies on the same basic principles of Boolean logic. Simply speaking, Boolean logic refers to using connectors between terms to define the search's parameters:

- *and* finds only those hits with both terms
- *or* finds any hits with either of the terms
- *not* excludes hits with the specified terms

As an example, suppose we are interested in writing a grant to fund development of an outpatient clinic in an urban setting; furthermore, we are interested specifically in a permanent clinic, not one using a van that visits local areas of the city. How might we use Boolean logic for an efficient search?

First, we might search on "grant" AND "clinic" as terms that both must appear. Depending on the response, we might need to try alternate parameters, such as "grant" OR "funding." We might also specify "urban" as the clinic type by using the compound search term "urban clinic." If hits appear for a mobile clinic, then we could try "urban clinic" NOT "mobile" to exclude those not part of the grant. (Check search engines for other allowed connectors beyond these three.)

In addition to Boolean logic, other techniques can enhance searches. In our clinic example, let's assume that "urban" as a term produces no hits. Although no funds might be available for urban clinics, a more plausible explanation may be that "urban" is not a term recognized by the search engine. We may need to check an internal thesaurus in the search engine for its preferred subject headings. Many healthcare databases, for instance, rely on *MeSH* (*Me*dical *S*ubject *H*eadings).[2]

Furthermore, engines allow advanced searches in which authors, dates, types of publications, and parts of publications can be selected. A search could focus on the years surrounding enactment of a particular piece of healthcare legislation. Searches can be limited to peer-reviewed journals with references. Even research articles vs. editorials can be differentiated. Again, spending time upfront to learn how an engine operates allows for a more efficient search while prewriting.

Effective Selection of Search Engines

We now know criteria for evaluating sources, principal types of information, and tips for efficient searching. One key remains to unlock the gate to healthcare information: proper selection of a search engine. A search engine is simply the device used to conduct a search. Examples range from online catalogs at local libraries to specialized databases with access fees. Beginning writers need to be able to differentiate these two main types of search engines:

- Generalized Internet search engines
- Specialized database search engines

Generalized engines refer to those with few access restrictions, such as those in a public library or linked through a user's web browser; in contrast, *specialized* engines use advanced techniques to access highly specialized databases, often for a fee. Generalized engines offer breadth across many topics, whereas specialized engines provide depth into a targeted field.

Writers, therefore, need to balance breadth and depth when choosing engines for a particular topic. A layperson brochure on health insurance may need just a superficial search, but a manual for administrators or practitioners requires more thorough coverage of governmental regulations. Table 5.3 provides examples of search engines that writers might consult. Each engine has its own advantages and disadvantages.

These examples demonstrate such characteristics (but again they do not endorse any one engine; other search engines not listed should also be considered).

Table 5.3 Generalized and Specialized Search Engines

TYPE OF SEARCH ENGINE	REPRESENTATIVE EXAMPLES
Generalized	Google, arguably the most omnipresent search engine, has an impressive breadth, although sources require more scrutiny of their reliability (Google Inc., Mountain View, CA)
	Google Scholar, a more academic version of the ubiquitous Google engine, allows searches across an extremely wide breadth of source materials (Google Inc., Mountain View, CA)
Specialized	MEDLINE provides in-depth technical information on medical topics; enhanced user-friendly engines, such as MedlinePlus, may be easier to search (National Library of Medicine, Washington, DC)
	CINAHL specializes in healthcare information for nursing, physical therapy, and related fields; it has an index of over 1,700 journal titles in healthcare (EBSCO Publishing, Ipswich, MA)

Let's walk through a search to compare and contrast two search engines: Google Scholar as a generalized engine emphasizing breadth, and MEDLINE as a specialized engine emphasizing depth. The topic is the recently approved vaccine for cervical cancer. Figure 5.1 summarizes hits derived from Boolean permutations using these two engines, employing advanced search techniques as well.

Figure 5.1 Comparison Searches on Generalized vs. Specialized Engines*

GOOGLE SCHOLAR (Generalized)	MEDLINE (Specialized)
Searched on "vaccine" AND "cervical" AND "cancer"	
about 79,000 hits	2,544 hits
Limited to 2013 sources	
5,570 hits	196 hits
Limited to subject heading "papillomarvirus vaccines"	
2,840 hits	51 hits
Limited to subject heading "adolescent"	
1,240 hits	28 hits

* *Results from searches conducted on the same day: October 10, 2013.*

Searching with the generalized engine (Google Scholar) amassed 40 times the number of hits that the specialized engine (MEDLINE) provided. In some cases, that breadth might be needed; however, assessing over 1,200 hits would be a daunting task. Given that 28 hits were still found with the specialized engine, the depth vs. breadth trade-off seems worthwhile. This example demonstrates how familiarity with search engines and advanced techniques can streamline later work in wading through voluminous sources.

Of course, writers would need to decide how to compile these references and any other sources. Too much material at hand—without proper organization—can sometimes be worse than not enough information. Experienced writers have their own ways of organizing material; one easy way, for example, is to note on the first page some of the key points, as well as how the information relates to the audience and purpose.

More formal ways include the annotated bibliography and literature review. In an annotated bibliography, the writer first compiles the reference list (bibliography), to which is added a summary or abstract for each entry; adding the relationship to audience and purpose can be useful. The literature review takes the annotated bibliography one step further. Unlike an annotated bibliography, a literature review avoids relegating each reference to its own paragraph. Instead, a text-based argument weaves together key points such as consensus, controversy, and other trends noted by researchers. While not required in all document formats, a literature review can provide a strong foundation for beginning sections of the document itself.

Substantiation of Document Content

Among the key steps that healthcare writers must take is citing sources for information used within documents. Substantiating content not only displays respect for others' work but also can protect writers from lawsuits regarding copyright infringement (at least if it is done correctly)—but *how* do writers substantiate sources?

Basically, writers need to cite references for any information they use that would not be considered common knowledge for someone in that field. In healthcare documents, for instance, writers do not need to use sources to substantiate that the aorta is the principal artery or that penicillin is an antibiotic. However, writers would need to cite references for the effects of undue pressure exerted by blood coursing through the aorta, as well as for rates of allergic reactions and anaphylactic shock attributed to penicillin.

Editorial conventions for references themselves, however, are beyond the scope of this book. Individual institutions, companies, and publishers may use their own "house-style" for references, although increasingly writers rely on common systems like those of the American Medical Association[3] (used in this book) or the International Committee of Medical Journal Editors.[4] Any standard referencing system, though, requires addressing the following key elements:

- **Author**—person, group, or organization responsible for the information; be especially careful of writers of chapters vs. editors of books
- **Title**—identifier for article, journal chapter, book, website, etc.; note that systems distinguish these by capitalization, italics, and quotation marks
- **Date**—time of formal publication for a piece, which may vary for parts of a full compendium; look out for original vs. revision dates on websites
- **Journal numbering**—identifiers to distinguish volumes and issues, as well as inclusive pagination; electronic pdf files give original pagination
- **Book publisher**—copyright owner for the document; include location (city, state, country, as needed), as well as imprints (i.e., subsidiaries)
- **Miscellaneous**—documents not fitting the molds of journals or books will need other data, such as ID numbers for government pamphlets

These elements for references apply no matter the document's mode of delivery: printed hard copy; electronic versions of hard-copy documents; or information that exists only in electronic format, such as a website. Sometimes, an unusual source needs to be adapted into the most appropriate format, particularly electronic sources not yet covered by available manuals. Besides giving credit for the information's originators, writers also need to credit the delivery system for documents obtained electronically. This credit falls into one of two categories:

- **Database sources**—search engines, whether general ones like Google or proprietary ones like PubMed or CINAHL (as already covered)
- **Website sources**—information either explicitly contained within that website, or documents made electronically available through that website

Begin by formatting the citation as if the source were obtained strictly as a traditional hard copy. Then add details that indicate the electronic means through which the source was obtained, using some variation of one of the following standard statements:

Available from: [name of database]. Accessed [date]. [database version]
Available at: [url of website]. Accessed [date]. [website version]

Note that both versions include the date of access. Remember that the primary purpose of substantiation is to allow readers to investigate further: because electronic information changes rapidly, the actual date of access can be informative. (In this book, the access dates have not been included for the sake of simplicity; access dates are more relevant for journal articles and websites, which are published more rapidly than are books.) Also note that database versions usually do not list the url (a lengthy search string) but rather the name of the database. (Some styles require a separate identification number for each document, which can be particularly important for references still in press.) In all cases, double-check specific statements with the selected style manual.

One last point that beginning writers may forget: substantiation of source material encompasses both direct verbatim quotation as well as indirect paraphrased rewordings. As writers, we are crediting the information itself, along with the way in which it was presented. Both aspects reflect the intellectual creativity of the originator. Soon, you'll be the original creator—won't you expect the rest of us to credit you?

Chapter Summary

- All potential information sources, especially those online, should be assessed for the main criteria of authority, accuracy, currency, and objectivity. Accessibility of sources is an insufficient justification.
- Online searches can be enhanced through careful use of Boolean connectors like *and*, *or*, and *not*. Advanced searching techniques expedite the search process by zeroing in on pertinent sources. Healthcare writers should be familiar with the *MeSH* (*Me*dical *S*ubject *H*eadings) thesaurus used in MEDLINE. Other databases may have different terminology sets.
- Search engines can be classified as *generalized* (breadth) or *specialized* (depth). Combined use of these two types of engines can be helpful, although specialized engines can focus the search process.

- The wide availability of document sources, particularly through electronic means, all require substantiation. Standard formats for referencing provide the specifics. Regardless, direct quotations and paraphrased rewordings require citations, since the underlying point is to credit someone else's idea.

Exercises for Practice

1. With healthcare sources so readily available, writers need to evaluate them carefully. Use the four criteria (authority, accuracy, currency, and objectivity) to assess the strengths and weaknesses of the following sources:
   a. promotional brochure from an insurance company
   b. college textbook borrowed from your professor
   c. scholarly article from a website with pop-up ads
   d. government flyer urging childhood vaccination
   e. package insert provided with a prescription drug
   f. television advertisement on a new drug treatment
   g. program summary issued by an advocacy group
   h. foreign reference book translated into English
   i. trifold brochure on a disease for a new patient
   j. multipage document with references and tables

2. Reliable sources for healthcare information can still be obtained from books and other documents in traditional hard-copy format. Use materials from a library, bookstore, or catalog to identify at least one additional hard-copy source for each of the groups in Table 5.1. Summarize the advantages and disadvantages of these additional sources.

3. Online sources for healthcare information are invaluable. Care must still be taken, though, because sites may have commercials or other slants. Choose a healthcare topic of interest to you; then conduct a search using online sources from sites listed in Table 5.2 (use these sites, or substitute your own). Do differences in the information that you find relate to the sites chosen?

4. Searches for information become more fruitful when writers know the tricks of effective searching. Identify how you could use Boolean connectors and advanced techniques in the following searches:
   a. diabetes that first affects adults later in their lives
   b. pros and cons of health maintenance organizations

c. current rates of malaria in sub-Saharan countries
d. healthy outcomes from diet, exercise, and lifestyle
e. requirements for pharmacists to maintain licensure

5. Even when writers know how to conduct searches, they need to select the most appropriate search engines for their topics. Use the healthcare topic from Exercise 5.3 to conduct a parallel search with two engines: one generalized and one specialized. (See Figure 5.1 for a model.) What do your search results suggest in terms of balancing breadth and depth when choosing search engines?

6. Identify which of the following pieces of information require substantiation as a cited reference. Find a source for each one, if you can:
a. Obesity is an increasingly common health problem in the United States.
b. The development of new drugs proceeds through four main phases.
c. Nurse practitioners can prescribe medications for certain indications.
d. Always consult your doctor before beginning a new exercise program.
e. The United States spends the most money on healthcare per person.

References

1 Reference Department, Wolfgram Memorial Library, Widener University. *How to Evaluate Information on the Web*. Chester, PA: Widener University; July 2011. http://www.widener.edu/about/campus_resources/wolfgram_library/evaluate.

2 US National Library of Medicine, National Institutes of Health. *Medical Subject Headings*. Bethesda, MD: US National Library of Medicine. First published September 1, 1999; last updated August 30, 2013. http://www.nlm.nih.gov/mesh.

3 American Medical Association Style Committee. *AMA Manual of Style: A Guide for Authors and Editors*. 10th ed. New York: Oxford University Press; 2007.

4 International Committee of Medical Journal Editors. *Recommendations for the Conduct, Reporting, Editing and Publication of Scholarly Work in Medical Journals (ICMJE Recommendations)*. Philadelphia: ICMJE Secretariat, American College of Physicians; 2013. http://www.icmje.org/recommendations.

6

EVIDENCE THROUGH TEXT ARGUMENTATION

Chapter Objectives

In this chapter, you will learn to:

- Construct evidence through classical rhetoric with persuasion
- Identify situations for using inductive vs. deductive reasoning
- Distinguish cause-effect relationships from temporal fallacies

Evidence as Building Blocks

After collecting information to populate the framework of strategies, writers next need to determine ways to use that information. In other words, what would be the most effective ways—for the intended audience and purpose—to arrange these building blocks within the document's outline? Writers tackle this writing phase by composing text and, as will be seen in the next chapter, preparing visuals.

With the full dictionary (including medical dictionaries) for their lexicon, writers have the opportunity to compose text, much like orators do with speech. Information can be packaged as word arguments—not disputes in the vernacular sense but rather elements of traditional debating. Beginning writers should be aware of three basic techniques (as listed in Table 6.1):

- *Classical rhetoric* is argumentation traditionally used in debating

- *Induction and deduction* link general theories with specific cases
- *Cause and effect* identify special cases of temporal relationships

Table 6.1 Descriptions and Examples of Word-Based Evidence Techniques

TECHNIQUE	DESCRIPTION	EXAMPLE
Classical rhetoric		
Logos	Uses numbers, statistics, source materials, and other "hard" evidence as backup to document	Statistically significant results from clinical trial of new drug.
Pathos	Uses anecdotes, personal accounts, or other "soft" evidence to evoke sympathy or empathy	News story on child with cleft palate to solicit contributions.
Ethos	Balances evidence from various perspectives so as to show fairness and unbiased approach	Different insurance plans now available to firm's employees.
Induction and deduction	Evaluates from cases to theorem (induction) or from theorem to cases (deduction)	HIV transmitted by transfusion. Stem cells can generate tissues.
Cause and effect	Reasons forward (if x occurs, then will y too?), or backward (if y occurs, did earlier x cause it?)	Does vitamin C prevent colds? Is dementia linked to plaques?

Classical Rhetoric for Persuasion

Today's healthcare audiences can still be persuaded through the classical rhetoric attributed to Aristotle during the fourth century BCE in ancient Greece. He identified three modes of persuasion that use the spoken word, as in debating, but these techniques can also be applied to written documents. Note that rhetoric formally applies to a particular purpose: to persuade. Nonetheless, writers can use Aristotle's three modes in documents with other explicit purposes because an element of persuasion underpins situations in which readers assess information.

In healthcare documents, then, writers can choose if classical rhetoric can be used to persuade a target audience. Aristotle identified these three modes of persuasion, as listed in Table 6.1:

- *Logos*—logic of evidence with forceful presentation
- *Pathos*—empathy fostered through emotional detail

- *Ethos*—credibility reinforced by remaining unbiased

Some audiences may be receptive to quantitative logic, such as statistical data, whereas others may be more receptive to emotional details, such as an anecdote. Mixing the two approaches might backfire with audiences attuned more to one approach than the other. Nevertheless, today's astute audiences will not fall for overly slanted views; hence, writers must maintain a sufficient level of credibility at all times.

Starting with logos, writers interested in healthcare already should be reasonably comfortable with healthcare evidence relating to anatomical facts and medical diagnoses. Logos, in fact, applies in most—if not all—healthcare documents at least at some level. Even a descriptive piece on diabetes written for a newly diagnosed child would include facts about this increasingly common malady, albeit written with language that remains informative but not condescending.

In that same example, writers would identify during prewriting that a child might be alarmed by a document that is too stolidly fact-laden. One approach for the writer would be to apply pathos. The document could open with an anecdote about a boy or girl experiencing common symptoms of diabetes, such as thirst. The child reader could then relate to the child in the story. Moreover, the document could use this anecdote as a unifying theme, following this child throughout the document: experiencing symptoms, being diagnosed, and coping with treatments. Pathos, then, would be a unifying device in a document that still provides considerable medical details through logos.

As for ethos, writers would strive for a balance that sufficiently assuages undue fears without diluting the necessary facts. Such an interjection of ethos underpins almost all healthcare writing. More important would be the writer's duty to protect the document from excessive external influence. Consider a writer for a biologics firm that markets an insulin product. Not all patients with diabetes require insulin. In this case, the secondary audience would be the child's parents who discuss the booklet with their child. Adults, however, are also susceptible to subtle coercion, as in subliminal messaging buried under the surface of advertisements. Again, ethos is critical.

Figure 6.1 provides excerpts from a 99-page report on caregiving for veterans.[1] Even this short selection exemplifies how writers can achieve an audience-appropriate balance of logos, pathos, and ethos. This report, from a study conducted by the National Alliance for Caregiving, funded by the United Health Foundation, is also made available through the website of the Administration on Aging of the US Department of Health and Human Services. This excerpt summarizes caregiving data, specifically if veterans have help (paid or unpaid) besides the principal caregiver.

Figure 6.1 Rhetorical Appeals in Describing Caregiving for Veterans

Presence of Other Unpaid and Paid Caregivers

Two-thirds of caregivers report that the veteran they care for has not received any care from paid caregivers (67%). Even unpaid help is not very common. Only one-quarter of caregivers indicate that the veteran has had at least a moderate amount of care given by other unpaid family members or friends (25%).

Those who are fortunate enough to have family members who help with the caregiving responsibilities typically name their own siblings, parents, or adult children as the people who help. They tend to provide periodic relief rather than substantial ongoing caregiving.

When no other family members provide a significant amount of help to the caregiver, it tends to be because they live out of town or are busy with jobs and children.

Occasionally, I'll have one of our sons come in from out of town and stay for a week and let me go away to my cousins' in Florida and do nothing. —Caregiver #11

My son and his wife both work full time and part time, and anytime I need their help and ask for their help they will do it. But as far as them going to do Annette's therapies and stuff, they don't have the time. They're always here when I bring Annette home, and they always stay until I get her back in the van. If I'm not around and she needs something, they'll pick it up for her and take it to her. It's hard enough to raise a family and work full and part time and do all the chasing you do with three kids.—Caregiver #4

The report highlights percentages (logos) for such care; not shown in this excerpt are bullets detailing statistics. While this text hints at pathos in words like "fortunate" to describe shared caregiving, a more obvious use of pathos occurs in call-outs as sidebars to the text (appearing throughout the report). These call-outs provide comments from caregivers interviewed for the study. The text also blends in ethos, as exemplified in the last paragraph. Easily, the text could have been slanted against family members who do not share caregiving duties; instead, explanations for their unavailability lets readers know that many such family members have little choice.

Induction and Deduction

Medicine, like any science, relies on connecting observable facts with theorized conclusions. As researchers accumulate more and more data in experiments, for example, they can eventually draw conclusions. Moreover, sometimes they theorize a relationship to be present, and then the researchers look for evidence to that effect. These two ways of looking at science form the basis of the next building blocks for healthcare documents: induction and deduction.

Induction depends on inducing, not in the sense of causing a response, such as inducing a coma to protect a brain-injured person, but rather inferring something more general from particular pieces of information. Suppose that oncologists (tumor specialists) begin noticing that a new drug for shrinking tumors in breast cancer works much better in women who have never smoked. At first, observations might be too sporadic to be scientifically meaningful. However, if this trend continues, eventually researchers may conclude that smoking appears to interfere with the drug's effectiveness for treating breast cancer. That, in essence, typifies induction: moving at some stage from individual observations (responses of non-smokers) to a broader conclusion (interference with the drug's action). Induction can be remembered by association with another "I": Isaac Newton; in the often-retold anecdote, the continuous falling of apples led to a "eureka" moment that gravity as a physical force causes objects to fall down.

Deduction, on the other hand, depends on deducing—applying a theorem to see if it works in real circumstances. Deduction can be remembered through association with the "D" of detectives like Sherlock Holmes. Postulating that villains have massive egos, the detective might set a trap with a diamond touted as being unstealable and therefore irresistible to an egomaniac. Theorems do not arise from fiction, though. Researchers might theorize that hormonally active drugs would be more effective against tumors in tissues like breast with endocrine receptors. In fact, drugs for treating breast cancer are categorized as estrogen-receptor positive or negative. That is an example of deduction: if breast tissue has endocrine receptors, then breast cancer might respond to hormones. As actual evidence has shown, some tumors respond but others do not, depending on the presence or absence of receptors for the female hormone estrogen.

Induction and deduction often appear in summaries of research for non-researcher audiences. Figure 6.2, for instance, provides two excerpts from an overview of research that eventually allowed gastric ulcers to be successfully treated with antibiotics—earning a Nobel Prize in Medicine for Dr. Robert Warren and Dr. Barry Marshall.[2] Warren began unexpectedly detecting the spiral-shaped bacterium *Helicobacter pylori* in the mucosal linings of many individuals with gastric ulcers. Because bacteria should not have been able to survive in the stomach's intensely acidic environment, prior researchers dismissed the bacteria as being due to contamination of samples of the stomach lining. Not Warren. After repeated observations of spiral bacteria in gastric biopsies, he induced "a pattern. There was a definite correlation between active, chronic gastritis and the presence of these bacteria."

Figure 6.2 Inductive and Deductive Reasoning to Link Bacterial Infections and Peptic Ulcers

INDUCTIVE REASONING

The story begins with Dr. J. Robin Warren. Dr. Warren was a pathologist at the Royal Perth Hospital (RPH) in Western Australia.... In 1979, after examining a hematoxylin- and eosin-stained section of a stomach biopsy from a man with severe gastritis, Warren noticed a thin blue line on the surface of the tissue. When he increased magnification, he thought he saw bacteria....

He began to see a pattern. There was a definite correlation between active, chronic gastritis and the presence of these bacteria. He also observed that the number of bacteria seemed to correlate with the degree of inflammation of the stomach lining—the more severe the inflammation, the more abundant the bacteria. But this didn't make sense. Medicine taught that the stomach was sterile. It was too acidic for any bacteria to survive for very long. Bacteria in the stomach had been reported before, but dismissed as a contaminant or as secondary to the problem. Warren, however, didn't believe his spiral bacteria were simply a contaminant.

DEDUCTIVE REASONING

In 1979, Dr. Barry Marshall moved to RPH to begin three years of internal medicine training.... The chief of gastroenterology suggested that Marshall work with Dr. Warren, investigating his gastric spiral bacteria. After speaking with Dr. Warren, Marshall's interest was piqued.

Marshall continued reviewing the literature for any information pertaining to their research. He read that while acid-reducing drugs relieved the symptoms of the ulcers, they did not cure them. The relapse rate was high. Cimetidine, a common acid-reducing agent, helped to heal the ulceration, but did not cure the disease. Marshall was further intrigued when he read that colloidal bismuth subcitrate (CBS; similar to the over-the-counter drug PeptoBismol) greatly reduced the relapse rate in those with gastric ulcers. **Marshall hypothesized that if a drug lowered the relapse rate for gastric ulcers and if bacteria can cause gastric ulcers, then CBS should have anti-microbial properties.** To test this, Marshall soaked a filter paper disc in CBS and then placed it in the middle of a Petri dish inoculated with *Helicobacter pylori*. A few days later, much to his delight, there was a clear zone of inhibition around the paper disc while cimetidine had no effect. This provided further evidence supporting the bacterial theory of ulcers.

Note: Bolded emphasis added.

Eventually, a connection was postulated, with the relationship later established definitively through deductive reasoning. Given the relationship between gastric ulcers and *H. pylori*, eradication of these bacteria should relieve inflammation in the mucosal lining that can lead to ulceration. Marshall deduced that, "if a drug lowered the relapse rate for gastric ulcers and if bacteria can cause gastric ulcers, then CBS [drug treatment] should have anti-microbial properties." Experimental evidence supported his theorem, with subsequent research proving the efficacy of treating gastric ulcers with antimicrobial agents—an unexpected conclusion at that time.

Cause and Effect

Regarding medical treatments, one could assume that a response was caused by a therapy if a patient improves afterwards. Such reasoning seems consistent with induction and deduction. Conclusions of causality, however, cannot be drawn simply from temporal (i.e., time-based) occurrences. We need evidence or proof of causation.

Consider depression. This mental illness is so much more than feeling "the blues" for a day or two. Severe depression can debilitate a person's daily life, obviating abilities to interact with others, be gainfully employed, or simply enjoy life. Someone suspected of being depressed may be prescribed a medication and asked to participate in counseling sessions. If, after one or two months of both interventions, the patient begins to feel much more like herself, then did drug therapy cause the improvement? Or counseling? Or both because neither was sufficient alone? Could another factor be at play? Perhaps friends became more supportive as they learned about depression. Evidence, presumably with statistical rigor, is needed to determine causality.

Often, causality is so tantalizing that individuals neglect to establish proof. In fact, a common problem in arguments of causality is termed the *post hoc fallacy*, derived from the Latin *post hoc ergo propter hoc* ("after this because of this"). In other words, because event #1 precedes event #2 does not signify that event #1 *caused* event #2. Establishing a link of causality necessitates sufficient evidence, and this evidence can be assessed in one of two manners: reasoning *forward* or reasoning *backward*. Reasoning forward asks how a given intervention could have led to observed changes; reasoning backward ferrets out what led to the current situation.

Obesity can be used to illustrate both approaches for causality. Reasoning forward might examine if public-health campaigns featuring celebrity athletes would lower body mass indices among teens, whereas reasoning backward might look at hours spent on the Internet. In either case, the reasoning depends on rigorous evidence that shows causality. Sometimes, the causative agent is a third factor: might teens eat more snack foods while watching sports events and surfing online?

Figure 6.3 provides examples of both types of cause-and-effect reasoning, as related to the contention surrounding childhood immunizations.[3] A published (but later retracted) report implicated childhood immunization as potentially causative for autism. Of course, many parents were alarmed, with some opting to forego immunizations for their children. Unfortunately, that decision placed their children at increased risk for measles, mumps, and other illnesses; moreover, those same children could spread any infections to others not yet immunized. In the excerpts, a physician addresses the concerns often expressed by parents. Reasoning backward, the physician explains that recent increases in measles can be traced back to decreased immunization. Reasoning forward, she extrapolates the risks for future outbreaks associated with non-immunization.

Figure 6.3 Relationships of Cause and Effect for the Childhood Immunization Controversy

... [H]ealth officials are seeing alarming rises in preventable diseases. The Centers for Disease Control and Prevention (CDC) has reported hundreds of US measles cases in 2011, the largest number in 15 years. Most of these occurred in people who were not immunized against measles.

Much of the controversy about vaccines stems from the now-debunked 1998 study that tried to link autism to the MMR vaccine that protects against measles, mumps, and rubella (German measles). Study after study has found no scientific evidence that autism is caused by any single vaccine, combination vaccines (like the MMR vaccine), or the mercury-containing preservative thimerosal, which was once widely used in many childhood vaccines but has since been eliminated.

Indeed, the journal that originally published the 1998 study retracted it and called the findings "a deliberate fraud." And the doctor behind the study lost his license. But the study and the attention it received influenced parents worldwide and contributed to a decrease in immunization rates. Indeed, recent polls indicate that 1 in 4 parents still think vaccines are linked to autism.

Some parents wonder why their kids need immunizations if many of the diseases they protect against are no longer commonly seen in the United States. But the fact is that infectious diseases that are rare or nonexistent here (because of immunization programs) are still huge problems in other parts of the world.

If immunization rates drop among US kids, an outbreak could be an airplane flight away if a disease is introduced by just one unimmunized person....

It's also important to understand the concept of "community immunity" (or "herd immunity").... A single person's chance of catching a disease is low if everyone else is immunized. But each person who isn't immunized gives a highly contagious disease one more chance to spread.

People who can't receive certain vaccines (such as infants, pregnant women, and those with compromised immune systems) are also protected when most of the population is immunized. So when parents decide not to vaccinate their kids, they not only put them at risk, but also others who cannot be vaccinated.

Note also the physician's awareness of her audience. The parents, as laypersons, would be emotional. Information had to be presented convincingly (logos) but without alarming them further (pathos). Moreover, she deftly covered considerations related to various aspects of the controversy (ethos), yet she still reinforced her point convincingly. Attention to ethos was especially important, as the sponsor agency (a hospital-affiliated foundation) might be expected to have a vested interest in promoting vaccination. Thus, writers must remain informed about not only the scientific aspects but also the emotional impacts associated with healthcare issues.

This chapter has exemplified many of the prewriting decisions that facilitate the writing phase through text argumentation. The next chapter explores the use of visual displays to complement such text.

Chapter Summary

- Logos, pathos, and ethos are the rhetorical foundations of classical argumentation in persuasion. Even for documents whose main purpose is not persuasion, these text techniques can guide construction during the writing phase.
- Researchers typically employ inductive and deductive reasoning to connect data with interpretation. These two forms of reasoning also facilitate explanations of scientific information to non-researcher audiences.
- Causality can be a difficult issue to address. Writers must avoid the fallacy that assumes causality for temporal occurrences. Reasoning forward and backward are two approaches for identifying a potential cause-and-effect relationship.

Exercises for Practice

1. This chapter introduced a number of text techniques that can be used to convey information efficiently and effectively to audiences. Identify the technique(s) relevant to the following healthcare situations:
    a. Booklet on nutrition designed for school-age children to help them make healthy choices among options in the cafeteria.
    b. Identification of the link between sickle-cell anemia and malaria for Peace Corps volunteers soon serving in sub-Saharan Africa.
    c. Expected effects of covering gym memberships for employees of a major corporation seeking to lower its healthcare expenditures.

(Note: You may want to research any topics that seem unfamiliar.)

2. One overlooked source of healthcare information is the reliable encyclopedia. Consider this encyclopedia definition for magnetic resonance imaging (MRI): "a noninvasive diagnostic technique that produces computerized images of internal body tissues and is based on nuclear magnetic resonance of atoms within the body induced by the application of radio waves."[4] Which

rhetorical appeal dominates? How could pathos be used in an expanded version targeting patients who will require an MRI scan? Does ethos come into play?

3. Like all medical interventions, vaccines have risks that must be considered against their benefits. Deleterious outcomes such as anaphylaxis are termed "side effects" if causally linked with evidence. The excerpt in Figure 6.4 recounts methods used to examine relationships between vaccines and adverse events.[5] How thoroughly did these researchers explore causality? The researcher committee categorized relationships by the level of available evidence. Using the referenced link, identify techniques of argumentation used in the report brief.

Figure 6.4 Examination of Vaccines for Side Effects Due to Causality

The committee considered the weights of evidence and then reached a conclusion about the causal relationship between each vaccine and adverse health problem pairing. The committee began from a position of neutrality, presuming neither causation nor lack of causation, and moved from that position only when the combination of evidence suggested a more definitive assessment regarding causation....

Based on the totality of the evidence, the committee assigned each relationship to one of four categories of causation in which the evidence:

- convincingly supports a causal relationship;
- favors acceptance of a causal relationship;
- favors rejection of a causal relationship; or
- is inadequate to accept or reject a causal relationship.

The committee did not use a category to designate evidence that convincingly supports no causal relationship, because it is virtually impossible to prove the absence of a very rare relationship with the same certainty that is possible to establish the presence of one.

References

1 National Alliance for Caregiving. *Caregivers of Veterans—Serving on the Homefront.* Bethesda, MD: National Alliance for Caregiving; November 2010. http://www.caregiving.org/research.

2 Meuler DA. *Helicobacter pylori and the Bacterial Theory of Ulcers.* Buffalo, NY: National Center for Case Study Teaching in Science, State University of New York, University at Buffalo; February 18, 2011. http://sciencecases.lib.buffalo.edu/cs/files/peptic_ulcer.pdf.

3 Ben-Joseph EP, reviewer. *The Risks of Postponing or Avoiding Vaccinations.* Wilmington, DE: KidsHealth, reviewed December 2011. Reprinted with permission. http://kidshealth.org/parent/positive.

4 "Magnetic Resonance Imaging." *MedlinePlus® Medical Dictionary*. Springfield, MA: Merriam-Webster, Inc. October 16, 2013. http://www.merriam-webster.com/medlineplus.

5 Institute of Medicine of the National Academies. *Adverse Effects of Vaccines: Evidence and Causality.* [Report Brief.] Washington, DC: National Academy of Sciences. Last updated September 4, 2013. http://www.iom.edu/Reports/2011/adverse-effects-of-vaccines-evidence-and-causality.aspx.

7

EVIDENCE THROUGH VISUAL DISPLAYS

Chapter Objectives

In this chapter, you will learn to:

- Provide numerical or verbal data within tables for easier access
- Select meaningful data relationships to highlight within figures
- Construct tables and figures to complement document wording

Evidence Built with Graphics

In addition to building evidence through words, as shown in the previous chapter, writers can also build evidence through graphics. Such visual presentations are especially helpful in conveying medical information. For example, to categorize all the drugs covered by a formulary or to describe the intricacies of the human body, a writer would need to amass many paragraphs of text. Without composing graphics, the writer would face a daunting task, but the task for readers would be even more convoluted.

As suggested in these two cases, graphics take one of two main formats:

- *Tables* that present individual data in rows and columns
- *Figures* that highlight spatial views or data relationships

Although tables come in many sizes and shapes, their typical rows-and-columns format remains constant. Figures, in contrast, can take diverse configurations (e.g., pie charts vs. pictographs). Regardless, all tables and figures serve to complement—not replace—text by assisting readers in digesting that information.

Since readers often refer to graphics without rereading the full accompanying text, tables and figures need to be understandable on their own. The following tips will assist beginning writers in getting the best use of tables and figures:

- Position the graphic as close as possible after the text that discusses it; a graphic not introduced in the text can easily confuse the readers.
- Provide a concise but informative title to distinguish this visual from any others; number tables and figures whenever multiples of either occur.
- Identify variables and units as headers above columns or on axes; convert data into forms that avoid excessive zeros or decimal places.
- Clarify unusual attributes (e.g., missing data or statistical significance) through footnotes; consider redesign if footnotes are excessive.
- Avoid unnecessary lines (especially vertical) that cause the visual to collapse in density; emphasize the information, not the visual itself.
- Credit references for information sources; reuse of data is like paraphrasing text, whereas directly copying would otherwise be plagiarism.

The following sections explore when and how to use visuals as building blocks to support the document framework. Details to be covered include how to decide when (and what kinds of) graphics should be used to present information according to audience and purpose, as well as some common pitfalls to avoid. (However, the technological creation of graphics using computer software will not be addressed; those procedures can be found online or in manuals with software packages.)

Tables for Data Compilation

All writers have seen, used, and prepared tables in some fashion; unfortunately, not all understand that structure stems from intended purpose. Tables efficiently provide data (whether numbers or words) in a format easily accessible in different ways. For example, a table listing incidences of commonly occurring cancers might subdivide rates by sex because some cancers (e.g., breast, prostate) occur more often in one over the other. Organization of rows and columns allows different readers to easily access rates by tissue site as well as by sex. Table 7.1 provides such cancer data.[1] Note its formatting:

- Title is concise yet descriptive, as well as numbered within chapter
- Headers identify variables, with rates in thousands to avoid zeroes
- Generous column spacing precludes excessive rulings (i.e., lines)
- Footnote to clarify that prostate cancer is not possible in women
- Reference noted within title (but could be provided as footnote)

Table 7.1 Estimates of Most Common Cancers in 2010 by Sex

CANCER SITE *(decreasing order)*	**ESTIMATED NEW CASES (IN THOUSANDS)**		
	TOTAL	**FEMALE**	**MALE**
Lung	223	106	117
Prostate	218	NA	218
Breast	209	207	2
Colon and rectum	143	70	72
Bladder	71	18	53
Melanoma of skin	68	29	39

Only the six most common cancer sites presented

NA = Not applicable

Note: Totals may not add up due to rounding.

As for text, information in tables is apportioned by writers so as to meet the audience and purpose. Laypersons and practitioners, for instance, may prefer accessibility over conciseness, whereas administrators referring to massive datasets might appreciate fewer pages. For example, the source version of Table 7.1 included additional data on less common cancers, as well as survival rates since 1980, thereby increasing the table's density by vertical rulings needed to differentiate tightly spaced columns. This choice of formatting represents a trade-off to be decided by the writer: are fewer total pages more desirable by the audience if those pages are not as easy to read?

Figures for Data Portrayals

If tables facilitate data access, then figures, by their overtly visual sense, highlight key trends and relationships or provide a spatial (i.e., dimensional) view not achievable in a table, let alone in text. For example, in Table 7.1, figures might be used to highlight the disparity in rates of bladder cancer in women vs. men, as well as the striking similarity in cancer of the colon and rectum between the two sexes. A diagram of a precancerous mole could be used to supplement a discussion of how to detect skin melanoma.

As many writers know, today's software allows any dataset to be portrayed in any of various figure options—some format choices, though, may lead to nonsensical figures. The design of figures depends on knowing which types of data work with which types of options. Regarding data types, numbers fall into three categories:

- *Continuous* data that form a number line
- *Ordinal* data that are ranked but not continuous
- *Categorical* data that are distinct but in no ranked order

Weight is an example of continuous data. Theoretically, values for weight could be as small or as large as physiology allows. Depending on the precision of instruments, measurements could also have as many decimal places as desired. In other words, a value could fall anywhere along a number line of weight. Categories of weight—underweight, normal weight, overweight, obese—are ordinal data, though. These categories can be ordered by size, but only selected limited categories are permissible. Finally, categorical data are not only non-continuous but also unranked. Tissue types (e.g., neural, muscular, skeletal) could be compared for the distribution of body weight; however, one tissue type is not of a higher rank than any other.

Data from the US Census Bureau will be used to examine formats and options for the different types of figures.

Line Graphs

Annual expenditures for personal healthcare[2] show two types of continuous data that can be represented as number lines: first, the years from 1960 to 2005 for the x-axis (horizontal); second, billions of dollars for the y-axis (vertical). Hence, these data lend themselves to line graphs (Figure 7.1). Note that, if only limited years were included in the dataset, then expenditures for those years would be considered ordinal in nature, thereby requiring a column graph instead.

Although the data reference provides annual expenditures, these line graphs were generated from every fifth year for simplicity. In both, axes are labeled; the y-axis labels include units, converted to billions of dollars to minimize insignificant digits. The titles concisely describe the contents, with a superscript pointing to the data reference (same for both line graphs). The two line graphs illustrate formatting options.

Figure 7.1 Ⓐ plots only one variable (total health expenditures), so no identifying legend is needed. The absence of gridlines keeps the graph less dense so that readers can quickly grasp the acceleration of expenditures since the 1980s. In contrast, Figure 7.1 Ⓑ compares total and personal health expenditures. The two plotted variables necessitate an identifying legend, as well as distinguishing styles/colors for the lines. Addition of major gridlines facilitates inspection of the two trends more closely. Actual values could be superimposed over each data point, if needed by the readers.

Figure 7.1 US Expenditures on Healthcare: 1960–2005

Ⓐ Simple One-Variable Line Graph

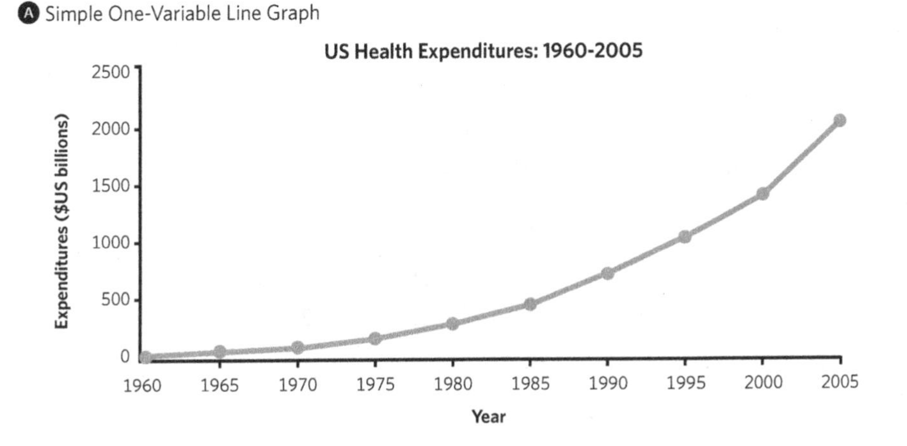

Ⓑ Two-Variable Line Graph with Legend and Gridlines

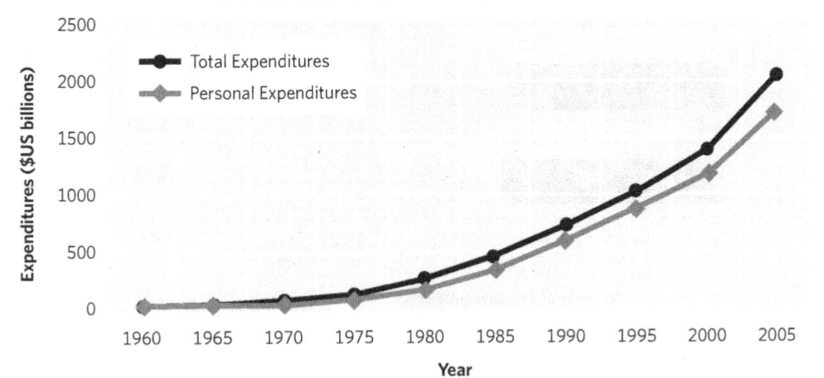

Column Graphs

For many datasets, one variable may be continuous but the other variable may not be. Figure 7.2 illustrates how column graphs can be used in these situations. The dataset[3] comprised the number of US persons according to type of healthcare coverage, limiting the comparison to the years 2008 and 2009. Unlike in Figure 7.1 for line graphs, time in Figure 7.2 is too restricted to be considered continuous; the fact that 2009 follows 2008, nonetheless, indicates that time should be presented ordinally. Categories of healthcare coverage immediately highlights that this variable is categorical; no overt ranking applies to one category over another.

Figure 7.2 Categories for US Healthcare Coverage: 2008 and 2009

Ⓐ Vertical Column Graph in Two Dimensions

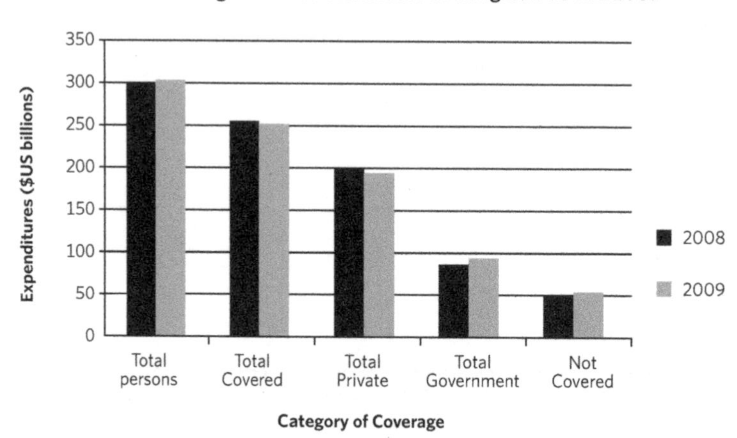

Ⓑ Horizontal Bar Graph in Three Dimensions

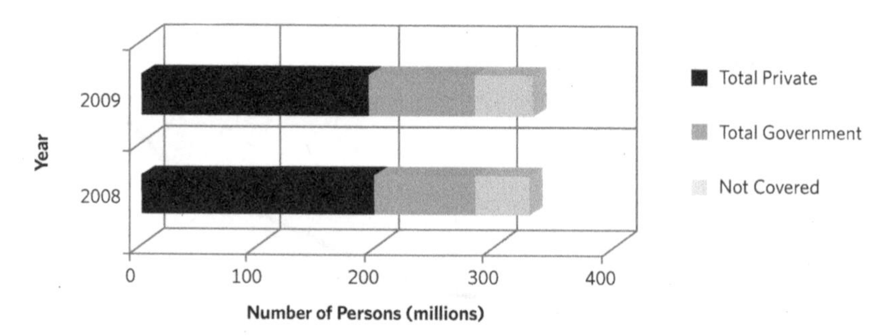

Figure 7.2 Ⓐ uses a vertical column graph to compare coverage (categorical) for numbers of persons (continuous) over two years (ordinal). The side-by-side comparison of 2008-2009 allows inspection of subtle changes. For example, although the number of total persons has slightly increased, the only coverage categories similarly increased are government coverage and no coverage. Notice the selective use of horizontal gridlines to allow interpolation from the graph. Coverage categories are identified on the horizontal axis; numbers in each category (in millions) are identified on the vertical axis; years are identified in the legend by color. (Patterns, termed "stippling," substitute for colors in documents printed in grayscale.)

Figure 7.2 Ⓑ shows a variation of analysis for the same dataset. In this case, the variables of interest are the categories of total private, total government, and not covered. These related categories have been "stacked" so that overall length indicates total persons without including that as a specific category. Juxtaposing the two stacks horizontally into a bar graph aligns these categories: for 2008 to 2009, total government and not covered have slightly increased, unlike total private. One other point about this variation is that this figure has been presented in three dimensions, an easy software trick. However, the use of three dimensions can render the data difficult to interpolate unless axes are also in three dimensions, as highlighted by the major gridlines.

Pie Charts

When datasets refer to the percentage distribution of parts within a whole, then a pie chart may be most useful for visual presentation. Pie charts simply and clearly allow subdivisions of a whole to be compared. For example, the distribution of physicians (e.g., family care, pediatrician, and so on) as a pie chart would allow determination of whether enough medical students are choosing non-specialist areas needed for efficient delivery of healthcare.

Notice, though, that pie charts can be overly restrictive as datasets become more complicated. First, all subdivisions must be included; categories not of specific interest cannot be ignored. As categories proliferate, smaller ones may need to be collapsed into an "other" category that may obfuscate analysis. Second, trends are not easily portrayed (as in column graphs). To represent changes over time, several pie charts would need to be placed side-by-side; more than two become clumsy. Third, information captured in a pie chart is restricted; hence, pie charts tend to work with administrators and laypersons more easily than with practitioners or researchers. For more knowledgeable audiences, distribution of categories might easily be captured in text.

Figure 7.3 illustrates the use of pie charts to examine the level of US households at risk for insufficient food (i.e., food insecurity).[4] The first pie chart Ⓐ illustrates that, in a prosperous nation like the United States, 15% of households were food at-risk in 2009. The title specifies the chart's content, the legend identifies categories, and the data labels provide actual percentages. The two-dimensional layout is unambiguous.

The second pie chart Ⓑ uses a similar dataset, this time the distribution of food insecurity in 2009 for those households with children. In this case, the percentage at risk is higher, at almost one-fourth of households. Note, though, that the slice seems smaller despite the greater percentage. This distortion results from rotating the three-dimensional chart. Of course, three-dimensional rotation could be applied in an unbiased manner. The point is that beginning writers must be careful with software tools—format options might inadvertently distort interpretation of the data.

Figure 7.3 Levels of Food Insecurity in US Households: 2009

Ⓐ Two-Dimensional Pie Chart without Rotation

US Households by Food Security: 2009

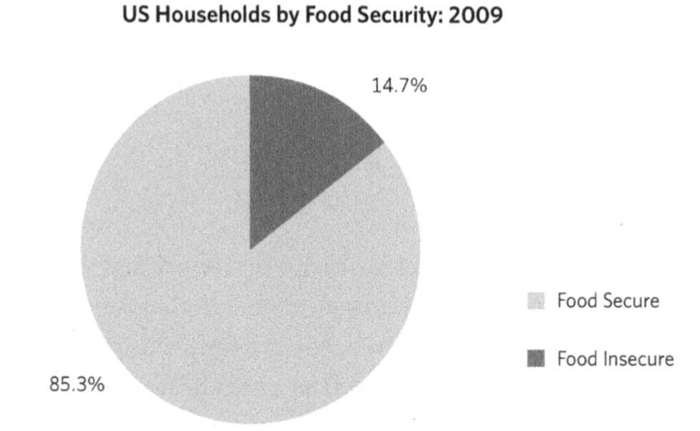

Ⓑ Three-Dimensional Pie Chart with Rotation

Food Security for US Households with Children: 2009

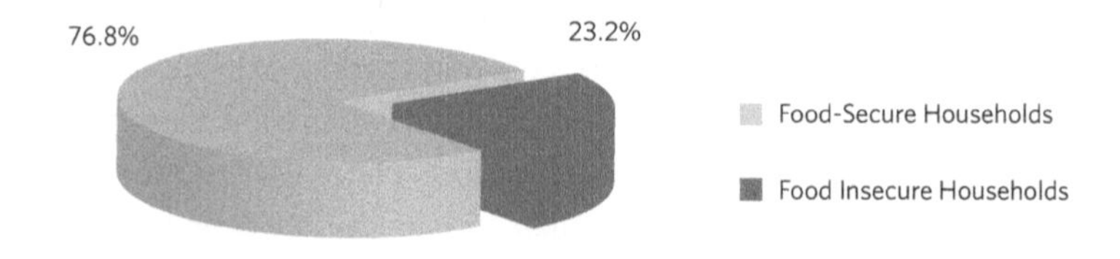

Other Visuals

Two other forms of graphics deserve attention by healthcare writers: pictorial depictions and flow charts. (Other types of visuals—from genome maps to CRT scans—are beyond the scope of this text.)

Pictures that show spatial relationships are especially applicable in healthcare: Figure 7.4 Ⓐ shows a stylized diagram of cancer cells in the skin, whereas Figure 7.4 Ⓑ uses an actual photograph of melanoma.[5] The first would assist layperson audiences to understand the underlying cellular structure of skin; the second allows comparison of a suspect mole to a true melanoma.

Flow charts, in contrast, show a progression through steps in a process. Examples in Figure 7.5 depict the same process for the development of a new pharmaceutical according to the guidelines of the US Food and Drug Administration (FDA). Figure 7.5 Ⓐ provides a simplified timeline within a presentation for administrators and scientists from the drug industry.[6] Although each step is clarified in additional slides, the writer could assume that this audience would know the basics, including undefined abbreviations. Figure 7.5 Ⓑ, in contrast, provides an eye-catching depiction of that process but for laypersons unfamiliar for the most part.[7] Interestingly, the visual for the layperson audience encompasses more information, an approach not typical for layperson-oriented documents. In this case, the astute choice reflects understanding of the target audience's interest in understanding this process in more detail.

Figure 7.4 Guidelines for Detection of Melanoma

Ⓐ Diagrammatic Representation of Skin Layer

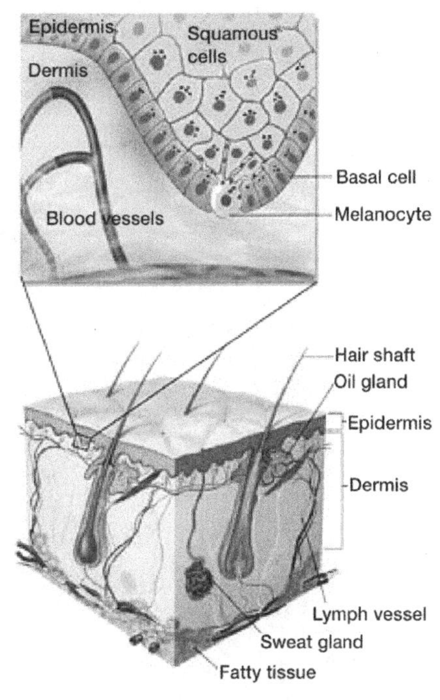

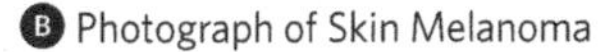

Ⓑ Photograph of Skin Melanoma

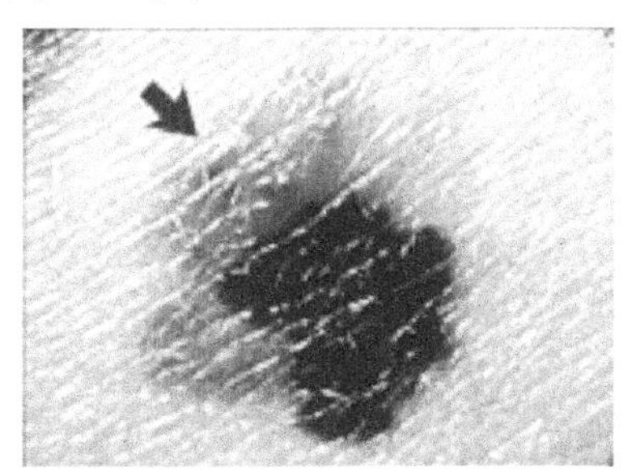

Figure 7.5 Process of New Drug Development

Ⓐ Flow Chart of Key Stages in Drug Development for Industry

Drug Development Timeline

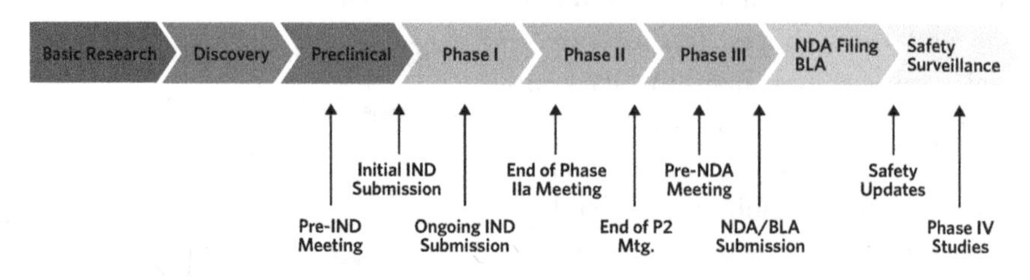

Ⓑ Infographic of Drug Development Process for Consumers

Additional Notes on Visuals

Depending on the target audience, figures can be modified to present more data or increase interest. Practitioners, for instance, may prefer that line graphs include statistical analysis (such as probability values), with standard error bars showing variability in mean points on the graph. Such complexity may not work for less knowledgeable laypersons or excessively busy managers; these two groups may appreciate column graphs or pie charts instead. Laypersons may enjoy the use of pictorial representations, such as stacked coins for healthcare expenditures.

Even beyond these various examples, writers have a plethora of choices. All that a writer needs is access to any graphics software to be overwhelmed by variety. Especially be cognizant of data types—software allows any data type to be plotted into any format, but no promises are made that the results will be meaningful. Finally, avoid options that might distort the data's meaning. Plotting graphs in three dimensions gives more visual impact but also can distort small differences if presented incorrectly. In short, the writer must choose from many possibilities to best meet the needs and preferences of the target audience. Just be careful not to write *down* to your reader—audiences of today tend to be more sophisticated about healthcare than ever before.

A key point too often forgotten by beginning writers is that all tables and figures need to be mentioned in the text (with the exception of bulky information appended at the back for completeness. Sometimes, a statement may simply point to the visual: "Table 14 lists mortality rates for men and women in the United States, from 1900 to 2000." Ideally, the pointing sentence provides more contextualization for the audience: "The greatest expenditures occurred for hospitalizations (see Table 22)." Some degree of interpretive commentary is important for charts indicating trends or relationships.

Handling text commentary, as well as text in general, relies on writing mechanics: the nuts and bolts of grammar, punctuation, diction, and so on. The next chapter explores the effective application of writing mechanics to meet the needs and preferences of target audiences for healthcare documents.

Chapter Summary

- Tables provide a rows-and-columns structure to organize data, whether numbers or words. This economical format applies when readers need or expect access to the actual data themselves.
- Charts and other figures allow writers to highlight notable trends and relationships within the data. Choice of format depends on the type of data: continuous, ordinal, or categorical. Software options should be used with caution.
- Tables and figures complement—but do not replace—the main text. If important enough to be included in a document, then a table or figure should be mentioned in the text so that readers are given context for assessment.

Exercises for Practice

1. Infection with Human Immunodeficiency Virus (HIV) remains a health concern across the globe. Rates of death attributed to HIV-related disease may be affected by demographic characteristics. For example, persons in certain age groups or of different ethnicities may be more at risk. Review the data provided in Table 7.2 (p. 97), which was taken directly from the US Census Bureau.[8] Can you suggest any ways to simplify this table? Write a short paragraph that could be in a document about the relationship between demographic characteristics and HIV death rates.

2. Despite sex education, teenagers often have children of their own. Birth rates among teenagers may be influenced by demographic factors. Review the data in Table 7.3 (p. 98) on births and birth rates among teenagers from 1990 to 2009.[9] Prepare at least two different types of charts to represent trends or relationships found in the data. Write a short paragraph for an accompanying report on this issue.

3. Software advances allow for increasingly varied formats for presenting data as visuals. The US Census Bureau, in fact, maintains a web-based gallery, from which Figure 7.6 (p. 99) was obtained.[10] This figure represents an unusual visualization of shifts in occupations; an added call-out highlights the healthcare occupations. Critique this visual (as well as any others linked through this referenced site). For which audiences would the format be most useful? How might you reconfigure this chart for simpler access to a wider group of readers?

Table 7.2 Death Rates from Human Immunodeficiency Virus (HIV) Disease by Selected Characteristics

[Rates per 100,000 population. Excludes deaths of nonresidents of the United States. Beginning 2000, deaths classified according to tenth revision of International Classification of Diseases.]

CHARACTERISTIC	1990	2000	2003	2004	2005	2006	2007
All ages, age adjusted[1]	**10.2**	**5.2**	**4.7**	**4.5**	**4.2**	**4.0**	**3.7**
All ages, crude[2]	**10.1**	**5.1**	**4.7**	**4.4**	**4.2**	**4.0**	**3.7**
Under 1 year	2.7	(B)	(B)	(B)	(B)	(B)	(B)
1 to 4 years	0.8	(B)	(B)	(B)	(B)	(B)	(B)
5 to 14 years	0.2	0.1	0.1	0.1	(B)	(B)	(B)
15 to 24 years	1.5	0.5	0.4	0.5	0.4	0.5	0.4
25 to 34 years	19.7	6.1	4.0	3.7	3.3	2.9	2.7
35 to 44 years	27.4	13.1	12.0	10.9	9.9	9.2	8.3
45 to 54 years	15.2	11.0	10.9	10.6	10.6	10.1	9.5
55 to 64 years	6.2	5.1	5.4	5.4	5.3	5.5	5.3
65 to 74 years	2.0	2.2	2.4	2.4	2.3	2.5	2.3
75 to 84 years	0.7	0.7	0.7	0.8	0.8	0.8	0.8
85 years and over	(B)	(B)	(B)	(B)	(B)	(B)	(B)
AGE-ADJUSTED RATES							
Male	18.5	7.9	7.1	6.6	6.2	5.9	5.4
Female	2.2	2.5	2.4	2.4	2.3	2.2	2.1
White male	15.7	4.6	4.2	3.8	3.6	3.4	3.1
Black male	46.3	35.1	31.3	29.2	28.2	26.3	24.5
American Indian, Alaska Native male	3.3	3.5	3.5	4.3	4.0	3.3	3.6
Asian, Pacific Islander male	4.3	1.2	1.1	1.2	1.0	1.1	0.8
Hispanic male[3]	28.8	10.6	9.2	8.2	7.5	7.0	6.3
Non-Hispanic, White male[3]	14.1	3.8	3.4	3.1	3.0	2.8	2.5
White female	1.1	1.0	0.9	0.9	0.8	0.7	0.7
Black female	10.1	13.2	12.8	13.0	12.0	12.2	11.3
American Indian, Alaska Native female	(B)	1.0	1.5	1.5	1.5	1.5	1.7
Asian, Pacific Islander female	(B)	0.2	(B)	(B)	(B)	(B)	(B)
Hispanic female[3]	3.8	2.9	2.7	2.4	1.9	1.9	1.8
Non-Hispanic, White female[3]	0.7	0.7	0.6	0.6	0.6	0.6	0.5

(B) Base figure too small to meet statistical standards. [1] Age-adjusted death rates were prepared using the direct method, in which age-specific death rates for a population of interest are applied to a standard population distributed by age. Age adjustment eliminates the differences in observed rates between points in time or among compared population groups that result from age differences in population composition. [2] The total number of deaths in a given time period divided by the total resident population as of July 1. [3] Persons of Hispanic origin may be any race. Excludes data from states lacking an Hispanic-origin item on their death certificates.

Source: U.S. National Center for Health Statistics, *Health, United States, 2009*. See also <http://www.cdc.gov/nchs/hus.htm>.

Table 7.3 Teenagers—Births and Birth Rates by Age, Race, and Hispanic Origin: 1990 to 2009

[Birth rates per 1,000 women in specified group. Based on race and Hispanic origin of mother.]

ITEM	NUMBER OF BIRTHS					BIRTH RATE				
	1990	2000	2005	2008	2009[1]	1990	2000	2005	2008	2009[1]
All races, 15 to 19 years	**521,826[2]**	**468,990**	**414,593**	**434,758**	**409,840**	**59.9**	**47.7**	**40.5**	**41.5**	**39.1**
15 to 17 years	183,327	157,209	133,191	135,664	124,256	37.5	26.9	21.4	21.7	20.1
18 to 19 years	338,499	311,781	281,402	299,094	285,584	88.6	78.1	69.9	70.6	66.2
White	354,482	333,013	295,265	306,402	(NA)	50.8	43.2	37.0	37.8	(NA)
Black	151,613	118,954	103,905	112,004	(NA)	112.8	77.4	62.0	63.4	(NA)
American Indian, Eskimo, Aleut	(NA)	8,055	7,807	8,815	8,316	81.1	58.3	52.7	58.4	55.5
Asian or Pacific Islander	(NA)	8,968	7,616	7,537	7,041	26.4	20.5	17.0	16.2	14.6
Hispanic[3]	(NA)	129,469	136,906	144,914	136,274	100.3	87.3	81.7	77.5	70.1
Non-Hispanic White	(NA)	204,056	165,005	168,684	159,526	42.5	32.6	25.9	26.7	25.6
Non-Hispanic Black	(NA)	116,019	96,813	104,559	98,425	116.2	79.2	60.9	62.8	59.0

NA Not available. [1] Preliminary data. [2] Includes races other than White and Black, not shown separately. [3] Persons of Hispanic origin may be any race.

Source: U.S. National Center for Health Statistics, National Vital Statistics Reports (NVSR), *Births: Final Data for 2008*, Vol. 59, No. 1, December 2010, and *Births: Preliminary Data for 2009*, Vol. 59, No. 3, December 2010.

Figure 7.6 Shifts in US Occupations by Sex: 2000 to 2006–2010

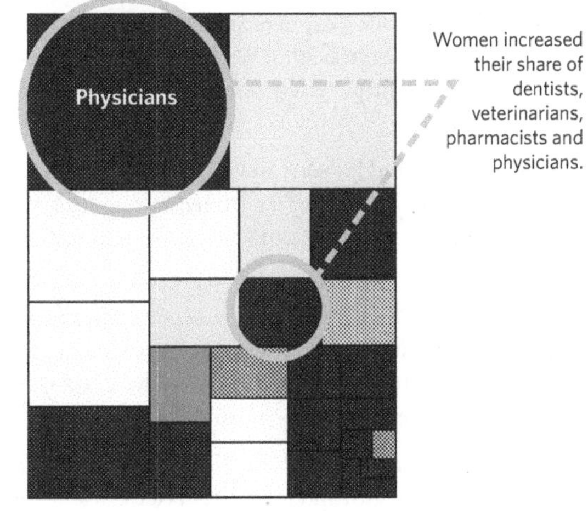
Deeper grey shades indicate gains in percent male.
Deeper dotted shades indicate gains in percent female.
Gains in percent male
No Change
Gains in percent female
Healthcare Occupations
Physicians
Women increased their share of dentists, veterinarians, pharmacists and physicians.

References

1 US Census Bureau. Cancer—Estimated New Cases, 2010, and Survival Rates, 1990-2007. [Table 182]. *Statistical Abstract of the United States: 2012*. Washington, DC: US Department of Commerce; last revised July 17, 2013. http://www.census.gov/prod/www/statistical_abstract.html.

2 US Census Bureau. National Health Expenditures—Summary: 1960 to 2009. [Table 134]. *Statistical Abstract of the United States: 2012*. Washington, DC: US Department of Commerce; last revised July 17, 2013. http://www.census.gov/prod/www/statistical_abstract.html.

3 US Census Bureau. Health Insurance Coverage Status by Selected Characteristics. [Table 155]. *Statistical Abstract of the United States: 2012*. Washington, DC: US Department of Commerce; last revised July 17, 2013. http://www.census.gov/prod/www/statistical_abstract.html.

4 US Census Bureau. Households and Persons Having Problems with Access to Food: 2005 to 2009. [Table 214]. *Statistical Abstract of the United States: 2012*. Washington, DC: US Department of Commerce; last revised July 17, 2013. http://www.census.gov/prod/www/statistical_abstract.html.

5 National Cancer Institute. *What You Need To Know About™ Melanoma and Other Skin Cancers*. Rockville, MD: National Institutes of Health, US Department of Health and Human Services; posted January 22, 2011. [Illustration © 2010 Winslow T; US Government has certain rights.] http://www.cancer.gov/cancertopics/wyntk/skin.

6 Yetter R. *PDUFA Activities in Drug Development*. Silver Spring, MD: US Food and Drug Administration, US Department of Health and Human Services; last updated January 5, 2011. http://www.fda.gov/downloads/ForIndustry/UserFees/PrescriptionDrugUserFee/UCM207568.pdf.

7 US Food and Drug Administration. *FDA Drug Approval Process Infographic*. Silver Spring, MD: US Food and Drug Administration, US Department of Health and Human Services; last updated March 3, 2012. http://www.fda.gov/Drugs/ResourcesForYou/Consumers/ucm295473.htm.

8 US Census Bureau. Death Rates From Human Immunodeficiency Virus (HIV) Diseases by Selected Characteristics. [Table 129]. *Statistical Abstract of the United States: 2012*. Washington, DC: US Department of Commerce; last revised July 17, 2013. http://www.census.gov/prod/www/statistical_abstract.html.

9 US Census Bureau. Teenagers—Births and Birth Rates by Age, Race, and Hispanic Origin: 1990 to 2009. [Table 84]. *Statistical Abstract of the United States: 2012*. Washington, DC: US Department of Commerce; last revised July 17, 2013. http://www.census.gov/prod/www/statistical_abstract.html.

10 US Census Bureau. Shifting Occupational Shares by Sex: 2000 to 2006-2010. *Data Visualization Gallery: A Weekly Exploration of Census Data*. Washington, DC: US Department of Commerce; posted July 15, 2013. http://www.census.gov/dataviz/visualizations/065/.

8

WRITING MECHANICS FOR HEALTHCARE

Chapter Objectives

In this chapter, you will learn to:

- Distinguish the major syntactic structures used in writing
- Match the sentence structures with their degrees of emphasis
- Avoid common pitfalls that can confuse target audiences

The Power of Writing Mechanics

After all the preparatory work during the prewriting phase, let's recap our achievements: audience, purpose, and context; medium, content, and strategy; sources; argumentation; and visuals. Writers at this point are fully prepared to use their outlines and notes to begin that long-awaited writing stage. As you begin that task—whether on computer or with pencil and paper—you must know how to string together your words into larger forms like sentences, paragraphs, sections, chapters, documents, and so on. Writers refer to the techniques for forming words into larger structures as "mechanics." This chapter opens the door for the writing phase. (Information on mechanics beyond this chapter's coverage can be found in many available writing guides.)

Before tensing up and risking writer's block, beginning writers should understand that mechanics are just another set of tools for writers to use when conveying information to audiences. Regardless of audience group, mechanics remain constant;

writers select the appropriate techniques, and then apply guidelines that give meaning to these selections. Mechanics are almost a psychological game: What can be done with words to emphasize or de-emphasize a point? Simplify a complicated concept? Encourage readers to continue rather than succumb to frustration? This is the power of mechanics.

Building Blocks of Syntax

Syntax—the structure of sentences—forms the foundation of writing mechanics. To understand syntax, writers must understand the substructures that compose sentences:

- *Word*: individual unit of writing
- *Phrase*: group of words without both subject (noun) and predicate (verb)
- *Clause*: group of words with both a subject (noun) and predicate (verb)
- *Sentence*: clause or clauses conveying a complete sense of meaning

Moving down this list from individual words to full sentences, each successive unit generally increases emphasis for selected information. A word, after all, conveys the least information, whereas a sentence conveys a complete meaningful thought and thus has the most potential for emphasis. (Of course, exceptions prove the rule: an individual word draws attention because of its rarity.) Adding another layer, clauses come in two varieties: independent clauses that make sense alone (much like regular sentences); and dependent clauses that, despite having a subject and predicate, do not quite make sense alone (and thus depend on other sentence structures to convey meaning). Some examples that could appear in healthcare documents comprise Table 8.1.

The building blocks in-between—phrases and clauses—require more explication. Phrases are simply groups of words that do not qualify as clauses because they lack *both* a subject (main noun and possibly adjectives) and a predicate (main verb together with possibly adjectives and direct or indirect objects). Phrases may have none or one but not both. Because phrases convey less information than do clauses, information for emphasis should not be relegated to phrases. Instead, phrases often carry descriptive or extraneous details not pivotal to the main idea of the sentence.

Table 8.1 Definitions and Examples of Syntactic Units

SYNTACTIC UNIT	DEFINITION	EXAMPLES
Word	Individual unit of meaning	Health; cancer; insurance; prostate; artery
Phrase	Group of words without both subject and predicate	[Note: Phrases are underlined with a dotted line] Lacking the training needed for this technique Without sufficient coverage for prescription drugs
Clause	Group of words with both subject and predicate	[Note: **Subjects** appear in bold; *predicates* appear in italics, dependent clauses are underlined once, and independent clauses are underlined twice]
Dependent clause	Does not make sense alone	Although **the appendix** *was removed* Whenever **a nurse practitioner** *prescribes therapy*
Independent clause	Does make sense alone	Given the urgency of the situation, **the medical resident** *quickly assigned persons injured in the traffic accident to the next available beds*.
Sentence	Word group with subject and predicate, and with full meaning	**The filled syringe** *must contain no air bubbles*. If **you** *experience any side effects*, **the pharmacist** *can consult with your doctor on alternative drugs*.

In some cases, phrases expand a noun into a nominative structure or a verb into a verbal structure (i.e., noun plus adjectives or verb plus adverbs, respectively). At other times, phrases may be more intricate by beginning with prepositions, participles, gerunds, or similar parts of speech. (Potential pitfalls of these types of phrases are addressed later in this chapter.) Just remember that phrases, lacking both a subject and a predicate, carry less emphasis: use them for ideas not critical for the target audience.

If a word group contains both a subject and predicate, then that structure qualifies as a clause. This distinction means more than nomenclature: deliberately use clauses to carry ideas or information that readers must know. However, clauses can be obfuscated in the distinction between dependent and independent structures.

As defined, any clause must contain both a subject and a predicate. Independent clauses function much like sentences: pare off extraneous structures, and the meaning remains. Dependent clauses, though, begin with a certain type of word that causes the information to feel incomplete; examples of such words are *although*, *among*, *because*, *between*, *if*, *since*, *whenever*, *whether*—to name a few. These words are grammatically identified as *subordinating conjunctions* because they render the clause *subordinate to* or *dependent upon* the main clause for meaning. If not for subordinating conjunctions, these dependent clauses could be stand-alone independent clauses.

To choose between phrases and clauses, writers must determine which information must be emphasized. If details are helpful but not necessary, then a phrase will suffice. Whenever information is crucial, though, a clause is needed; furthermore, independent clauses provide greater emphasis than dependent clauses do.

Figure 8.1 provides examples from a one-paragraph excerpt of the Foreword to a larger document on educational practices to manage food allergies.[1] The key information is conveyed through clauses, which can be read in sequence. Despite longer sentences in this overview, all its clauses are independent. (If an independent phrase is interrupted by a phrase, the two clausal sequences grammatically count as one clause.) The few phrases in the overview add the time frame ("In 2011"), shared responsibilities ("in consultation with"), and sentence transitions ("In response"). In this example, no situations required dependent clauses to show subordinate relationships.

Figure 8.1 Syntactic Units in Explanation of Food Allergy Guidelines

Font code: Phrase
Independent clause
Continuation of previous independent clause

In 2011, Congress passed the FDA Food Safety Modernization Act to improve food safety in the United States by shifting the focus from response to prevention. Section 112 of the act calls for the Secretary of US Department of Health and Human Services, in consultation with the Secretary of the U.S. Department of Education, to develop voluntary guidelines for schools and early childhood education programs to help them manage the risk of food allergies and severe allergic reactions in children. In response, the Centers for Disease Control and Prevention of the US Department of Health and Human Services, in consultation with the US Department of Education, developed the Voluntary Guidelines for Managing Food Allergies in Schools and Early Care and Education Programs.

Hierarchy of Sentence Structure

These building blocks of syntax can be combined in many ways. Focusing on clauses only (ignoring phrases, which do not affect syntactic class), writers can construct four types of sentence:

- *Simple:* only one independent clause
- *Compound:* two (or more) independent clauses

- *Complex:* one independent clause plus one (or more) dependent clause(s)
- *Compound-Complex:* two (or more) independent clauses plus one (or more) dependent clause(s)

Note that inclusion of phrases does NOT change a sentence's category; clauses are the sole determinant.

Table 8.2 provides examples of these four types of sentence. In these examples, the writers chose sentence types not only to match the audience but also to reinforce the information to be conveyed. As for matching the audience, a casual layperson might be distracted by the overuse of compound-complex structures in a simple educational set, whereas a more knowledgeable practitioner may need to grasp the intricate relationships that such a syntactic structure can convey. The examples in Table 8.2 show this increasing sophistication in syntax.

Table 8.2 Definitions and Examples of Sentence Types

Font code: Dependent clause vs. independent clause.

SYNTACTIC UNIT	DEFINITION	EXAMPLES
Simple	Only one independent clause	The physician leads the therapeutic team. At this healthcare facility, the family physician leads the therapeutic team.
Compound	Two (or more) independent clauses	At this healthcare facility, the family physician leads the therapeutic team, and the patient functions as her partner.
Complex	One independent clause plus one (or more) dependent clause(s)	Although the family physician leads the therapeutic team, the patient functions as her partner at this healthcare facility.
Compound-Complex	Two (or more) independent clauses plus one (or more) dependent clause(s)	Although the family physician leads the therapeutic team, the patient functions as her partner at this healthcare facility; moreover, the nurse, pharmacist, and therapist all play key roles.

Do not assume, though, that no compound-complex sentences should ever appear in documents for laypersons; rather, discriminating use is needed. The same tenet can apply to healthcare managers, who are busy. Practitioners (and researchers) more often need syntax beyond simple sentences to convey technical concepts. Nevertheless, short simple sentences can remain powerfully in any reader's mind.

Let's return to Table 8.2 to examine how syntax can reinforce information. The first example, a simple sentence, places the physician atop the therapeutic team. Although the second example is also simple, notice that the introductory phrase "at this healthcare facility" and the modifier "family" qualify this leadership role further. The next example, a compound sentence, builds on the previous qualified example to indicate the partnering role of the patient, which is critical in today's healthcare world. Structuring "the patient functions as her partner" as an independent clause reinforces a duality for the partners—physician and patient.

The example of a complex sentence takes this relationship another step further: subordinating the physician's role relative to the sentence's main clause on the patient's role highlights ongoing changes from a traditional paternalistic model. The final example adds the subtlety of these changing relationships by interjecting roles for other healthcare professionals. Structuring these additional roles within an independent clause underscores the mutability of such relationships.

The last point regarding syntax concerns grammar and punctuation. Dependent clauses require a comma when they open the sentence because they disrupt the normal flow of subject-predicate; dependent clauses following the main clause, though, are not disruptive and thus require no comma. Compare these two versions:

> "Even though cruciferous vegetables are healthy food choices, patients taking warfarin should avoid broccoli."

> "Patients taking warfarin should avoid broccoli even though cruciferous vegetables are healthy food choices."

Unlike complex structures, compound structures provide many nuanced options that convey particular information relationships. As listed below, compound structures linking independent clauses can use one of five main approaches:

, *coordinating conjunction*	comma plus one of seven coordinating conjunctions
;	semicolon (alone)
; *conjunctive adverb,*	semicolon followed by conjunctive adverb and comma
:	full colon
—	"em" dash (from the symbol font)

Writers choose among these five options according to the relationship that they desire to be conveyed to the audience. The plainest connection in a compound sentence uses a comma followed by a *coordinating conjunction*. English has only seven (no more, no less) coordinating conjunctions that can link two independent clauses with a comma. You can remember these seven coordinating conjunctions by the mnemonic (memory-aid) term FANBOYS, which gives the first letter of each of the seven coordinating conjunctions: **f**or, **a**nd, **n**or, **b**ut, **o**r, **y**et, **s**o. Joining two independent clauses with a comma only is termed a "comma splice" and should be avoided. (Some grammarians exclude overly short structures from the comma-splice rule.)

Despite (or perhaps because of) their power and omnipresence, these coordinating conjunctions do not provide much punch or nuance. One way to attract attention is to link two independent clauses by the second option of a solo semicolon. This structure causes readers to realize that the two ideas in these clauses are connected; they must intuit that connection (as exemplified in this very sentence). When not overdone, the solo semicolon can jolt the reader into paying closer attention.

Writers may not always desire to jolt readers; instead, writers may want to carve exact relationships between two clauses with subtlety and nuance, leaving no ambiguity for the reader (as exemplified in this sentence). In this case, after the *semicolon* come a *conjunctive adverb* and a comma. Conjunctive adverbs form a collection of transitional words: *furthermore, indeed, rather, however, nevertheless, moreover, hence, in addition,* to name a few. With so many options, conjunctive adverbs paint shades of meaning for relationships between independent clauses. Again, note the use of both a semicolon and a comma (not two commas) to surround a conjunctive adverb; two commas simply produce a more elegant comma splice. One caveat: balancing commas are used on either side of a conjunctive adverb that is interposed within a single clause, as in this example:

> "Drugs with limited marketing potential, however, may not be prioritized by pharmaceutical firms."

The remaining two options for joining two independent clauses into a compound sentence occur infrequently: a *full colon* (as used here) or an *em dash* (a longer dash the length of a lower-case letter "m" from the symbol font) can be used to emphasize how the second independent clause completes the thought of the first independent clause. These two options do crop up, particularly for lay audiences. When used too frequently, however, em dashes become stale—much like exclamation points! (The first sentence in this paragraph exemplifies the full colon, whereas the third sentence exemplifies the em dash.)

The rules of grammar and punctuation just reviewed for compound and complex sentences additionally apply to rarer instances of compound-complex sentences. Often, beginning writers shy away from this last structure, saving such esoterica for researchers. Although intimidating to some audiences, compound-complex sentences should remain in the writer's toolbox; in some situations, this structure is needed to capture the nuanced relationships in healthcare (or grammar, as in this sentence). Of course, writers should not pepper documents with complicated syntax merely to demonstrate their skills. Instead, these options should be selected based on audience and purpose.

As an example, the excerpt in Figure 8.2 comes again from the guidelines document on handling food allergies within educational settings.[2] This excerpt, written for administrators implementing these recommendations, is marked to distinguish independent and dependent clauses. Of the excerpt's 10 sentences, half (5) are simple, almost half (4) are complex, barely any (1) is compound, and none is compound-complex. The relatively high use of complex sentences emphasizes the interdependencies in treatment decisions. The short compound sentence highlights the fact that implementation of recommended actions is important—even when some actions appear more productive than others.

In the original document, this text overview was followed by a bulleted list of actions that could be implemented. Taken together, this overview and its ensuing actions fulfill the purpose of informing school practitioners about recommendations to safeguard the health of children with food allergies. Note that this larger document applies different writing mechanics (as well as strategies) in different sections. As discussed earlier, that technique is referred to as "nesting" strategies for smaller sections within an overarching structure for the document as a whole.

Figure 8.2 Clauses Used in Practitioner Guidelines on Food Allergies

Font code: Dependent clause vs. independent clause. [SENTENCE TYPE]

Effective management of food allergies in early care and education (ECE) programs requires the participation of many people. [SIMPLE] This section presents the actions that ECE program staff can take to implement the recommendations in Section 1. [SIMPLE] Some actions duplicate responsibilities required under applicable federal and state laws, including regulations, and policies. [SIMPLE] Although many responsibilities presented here are not required by statute, they can contribute to better management of food allergies in ECE programs. [COMPLEX]

If the ECE program participates in USDA's Child Nutrition Programs, the ECE program must follow USDA statutes, regulations, and guidance for providing meal accommodations for children with food allergy disabilities. [COMPLEX]

Some actions are intentionally repeated for different staff positions to ensure that critical actions are addressed even if a particular position does not exist in the ECE program. [COMPLEX] This duplication also reinforces the need for different staff members to work together to manage food allergies effectively. [SIMPLE] All actions are important, but some will have a greater effect than others. [COMPOUND] Ultimately, each ECE program must determine which actions are most practical and necessary to implement and who should be responsible for those actions. [SIMPLE]

Although these guidelines are specifically for licensed ECE programs, many of the recommendations can be used in unlicensed child care settings. [COMPLEX]

Avoidable Problems with Syntax

Given these various choices in syntax, grammar, and punctuation, problems can arise. Let's take a look at some of the more common problems that, once aware, writers can avoid. Some of these may be more easily detected during the postwriting or revision stage (which Chapter 12 covers), rather than while actually writing.

Restrictive Sense of Relative Clauses

Beyond the distinction between independent and dependent clauses, one further topic needs explication: the special case of relative clauses. By definition, relative clauses are dependent clauses starting with a relative pronoun—the words *that* and *which* in most situations; as well as *who, whom, whoever,* and *whomever* in situations regarding persons. The choice of relative pronoun, along with the inclusion (or not) of commas, depends on whether the relative clause is *restrictive* (i.e., critical to a sentence's meaning) or *non-restrictive* (i.e., not critical to a sentence's meaning).

For example, suppose that a doctor orders a set of laboratory tests to be run on a patient's blood sample to assist with diagnosis. Also suppose that the hospital's lab had recently purchased a new device with a controversially high price tag. Moreover, the lab results allow identification of a rare condition that would have been fatal to the patient if it had not been diagnosed early enough. Which of these two variations about the role of the new lab device would be correct? (Italicized text identifies relative clauses.)

"Results of tests *that were run on the new XYZ 2000* saved the patient."

"Results of tests, *which were run on the new XYZ 2000,* saved the patient."

Given an intent to spotlight the expensive device's value, writers would choose the first version: only the new XYZ 2000 was capable of detecting the rare abnormality. Hence, this relative clause is restrictive because its meaning is crucial to the sentence's main idea. In contrast, if the XYZ 2000 was the usual equipment used and all tests were routinely run on it, then identifying the device would be superfluous. This time, writers would choose the second version for its non-restrictive structure.

Figure 8.3 provides a short definition and description of enzyme inhibitors used in cancer therapy; this document originated from the American Cancer Society.[3] These two easy-to-read paragraphs contain four relative clauses (as highlighted). The first relative clause is non-restrictive: "... enzymes, *which* are special proteins...." This clause defines the term "enzymes" for readers who may be unfamiliar, but this definition is not essential. In contrast, the second clause is restrictive: "... are those *that* help digest...." This clause narrows down which enzymes of the many types are responsible for digestion; the clause is thus essential. Similarly, the third clause is also restrictive: "... enzymes *that* are signals for cancer cells...." This clause identifies the enzymes pertinent to this therapy. And just to make the situation a bit more tantalizing, the fourth relative clause is a restrictive clause embedded within a non-restrictive clause: "... proteins that help control...." This sub-clause is restrictive because it delineates the proteins with a controlling function. Albeit subtle, these distinctions convey meaning that writers may know but audiences need to grasp.

Figure 8.3 Relative Clauses to Define and Describe Enzyme Inhibitors

Font code: Restrictive clause vs. non-restrictive clause.
Restrictive clause embedded within non-restrictive clause.

Enzyme inhibitors

Our bodies produce many types of enzymes, which are special proteins that help control many of the things our cells do. When most people think of enzymes, the first ones that come to mind are those that help digest (break down) the food we eat. But some enzymes serve as signals for cancer cells to grow.

Some targeted therapies block (inhibit) enzymes that are signals for cancer cells to grow. These drugs are called *enzyme inhibitors*. Blocking these cell signals can keep the cancer from getting bigger and spreading. So even if the tumor is not getting smaller, its out-of-control growth has been interrupted. This may give regular chemo a better chance to work. Slowing or stopping out-of-control growth may also help people live longer, even without adding other drugs.

In this excerpt, note the distinction of restrictive vs. non-restrictive clauses (and their meanings) as captured by using *that* without commas (restrictive) vs. *which* with commas. While some grammarians interchange *that* and *which* as long as commas are used correctly, the selection of one relative pronoun over the other precludes confusion in the reader's mind. For relative clauses about persons, the commas are critical because both restrictive and non-restrictive structures should use the same term from the *who* family.

Consider the situation that follows:

> "Treatment histories can be provided by the operator of the clinic's computer, *who* is often a nurse practitioner but may be a nurse's aide."

In this non-restrictive relative clause, *who* (rather than *which*) refers to the operator of the clinic's computer. Interestingly, the *who* (*operator*) is modified by a prepositional phrase (*of the clinic's computer*). The use of *who* avoids any potential confusion, as could occur if the relative clause were restrictive:

> "Treatment histories can be provided by the operator of the clinic's computer *who* has been thoroughly tested for compliance with all regulations."

If *that* had been used, the reader might misconstrue that the *computer* had been tested for compliance (i.e., does the software follow necessary algorithms?), rather than the *person* running the computer (i.e., has the nurse's aide at the computer station completed training on compiling medical histories for patients at the clinic?). Rather than risk obfuscation, focus on the clarity that grammar affords to those who follow the rules.

Misplaced and Meaningless Phrases

Unlike clauses, phrases lack both a subject and a predicate. Not carrying as much information, phrases typically elaborate on information within an associated clause of the sentence. Depending on its structure, a phrase can function as an adjective to describe a subject, object, or other noun; or as an adverb to expand on a verb's action in a predicate. Two examples follow:

Adjective: With a dangerously low diastolic blood pressure, the patient almost lost consciousness whenever he tried to lift his head.

Adverb: The attending nurse quickly hooked up an intravenous line of saline in an attempt to raise the patient's blood pressure.

The two highlighted structures are both classified as "prepositional phrases" because they begin with a preposition (i.e., *with* and *in*, respectively). Typical for phrases that function adverbially, the prepositional phrase in the second example could appear in any of several positions within the sentence, without causing any confusion for the reader:

Option 1: In an attempt to raise the patient's blood pressure, the attending nurse quickly hooked up an intravenous line of saline.

Option 2: The attending nurse, in an attempt to raise the patient's blood pressure, quickly hooked up an intravenous line of saline.

Adverbial phrases tend to have that strength of clarity regardless of their position within the sentence. Certainly, some variations sound more natural than others;

any that sound remotely jarring should be avoided (unless the purpose is to get attention). Phrases functioning as adjectives, however, lack that clarity and thus can obfuscate meaning for some audiences. Therefore, writers should always position an adjectival phrase adjacent to the noun (or noun phrase) to be modified.

In the first example of an adjectival phrase, no confusion would occur regardless of position. Clearly, "with a dangerously low diastolic blood pressure" can modify only "the patient"; the other choices ("consciousness" or "head") would be non-sensible. Such is not always the case, though, as in this next example:

Clear: Weighing quality of life against sheer length, **the cancer patient** decided against aggressive chemotherapy as prescribed by the oncologist.

Unclear: Weighing quality of life against sheer length, **a decision** against aggressive chemotherapy as prescribed by the oncologist was made by the cancer patient.

In the clear version, "the cancer patient" weighs quality vs. length of life in opting against the aggressive chemotherapy favored by the oncologist. In the unclear version, "patient" is not only the last noun, but also the last word in the entire sentence. Grammatically, this sentence states that "a decision" weighs the two factors, which we know to be impossible. But couldn't both oncologist and patient be weighing these options? Or perhaps only the oncologist, leaving the patient to consider family budget or another factor?

The problem just exemplified represents a *misplaced modifier*: one not next to its *referent* (noun or pronoun being modified). This common problem arises with participial phrases (i.e., those that begin with a participle: the *-ed* or *-ing* form of a verb; *weighed* would be a past participle, whereas *weighing* would be the present participle). When the participial phrase functions adjectivally, the position of the phrase can be especially crucial. Consider the following permutation:

Dangling: Weighing quality of life against sheer length, a decision against aggressive chemotherapy was made.

This time, the participial phrase can modify either "decision" or "chemotherapy," neither of which makes sense. This sentence, in fact, contains no sensible noun for the participial phrase to modify. For this reason, the term *dangling participle* was coined—the participial phrase dangles in mid-air, most likely causing confusion.

Avoiding the potential confusion of misplaced modifiers or dangling participles comes down to two choices: (1) moving the phrase immediately adjacent to the intended referent (and adding that noun, if missing); or (2) expanding the phrase into a clause, as shown in this reworked version:

Correct: As the patient weighed her quality of life against sheer length, she decided against the aggressive chemotherapy prescribed by her oncologist.

Since all clauses contain both a subject and predicate, expanding the troublesome phrase into a dependent clause avoids the problem. Always remember that the writer's job is to use grammar to clarify the intended meaning for the target audience.

Active, Passive, and Nominalized Verbs

Often attracted by the subjects of sentences, readers may neglect the predicates—but that is grammatically where all the action occurs. Verbs in the predicates of sentences tell readers what actually happens. Hence, let's review these two grammatical aspects of verbs: voice and nominalization.

Voice refers to whether the verb is *active* or *passive*. With active verbs, the subjects perform the activity; with passive verbs, the subjects receive the activity instead. Consider these variations of the same sentence:

Active: The patient stopped therapy against the doctor's advice. [Who stopped therapy? The patient did.]

Passive: Therapy was stopped against the doctor's advice. [Who stopped therapy? No one is identified.]

Passive: Therapy was stopped by the patient against the doctor's advice. [Who stopped therapy? The patient did, but "therapy" has now been cast as the grammatical subject of the sentence, lessening the impact.]

Although many individuals tend to write with more passive constructions than needed, a predominance of active constructions keeps the readers' focus on the intended actions in the sentence.

Sometimes, though, writers may deliberately use passive voice if readers do not need to know (or simply would not care about) the actual subject, as below:

Passive: Finally, the long-term patient was discharged from the hospital. [Readers are more interested in the patient being discharged than in the clerk who processed the paperwork.]

In such instances, writers might intentionally select passive constructions. In most cases, however, writers should recast passive constructions into active ones. Writers often catch these constructions during the postwriting stage (which Chapter 12 covers).

As for *nominalization*, this term refers to turning an otherwise strong verb into a longer noun ending in *-tion*, *-ence*, and other such suffixes; typically, nominalization also requires moving a nominalized (i.e., made into a noun) verb into a prepositional phrase. In the process, the lengthened sentence often requires a passive construction:

Active: **Doctors without Borders** *vaccinated* the village children against outbreaks of typhus that often devastate developing countries.

Nominalized and Passive: **Vaccination** of the village children against outbreaks of typhus that often devastate developing countries *was done by* Doctors without Borders.

Notice that the nominalized version makes "vaccination" the focus, which may be the writer's intent. However, if the intent is to spotlight those physicians who care for the ill regardless of politics, then pull them up front to the subject and recast the nominalized "vaccination" into the active "vaccinated." As for voice, decide on nominalization on the basis of audience and purpose.

Pronouns, Expletives, and Other Foibles

This truncated review of mechanics in this chapter precludes an examination of all grammatical oddities; nevertheless, a few more merit attention. The first is the pronoun. These useful words (such as *I*, *her*, *you*, *they*) stand in lieu of nouns (termed *antecedents*). Pronouns obviate the overuse of nouns or noun phrases, which can result in long strings of nouns, particularly in some medical writing. For example, once establishing alprazolam as a treatment for anxiety, writers may then refer to the drug as *it* in later parts of the text. Trouble may arise, though, when the text covers more than one drug. Does *it* refer to alprazolam or to buspirone if both drugs are being considered for the patient's condition? Readers often assume that the antecedent is the more recently used noun, which may not be the intended case. Avoiding pronouns in vague circumstances can reduce such confusion.

As another issue, pronouns can unintentionally introduce sexism into writing—using *he* for a physician and *she* for a nurse, as examples. One way to avoid sexism is to pluralize, since plural pronouns (e.g., *they* or *their*) are gender-neutral. Writers must then be sure, however, that the pronoun's antecedent is also plural:

Sexist: The **nurse** always stands up for **her** patients.
Incorrect: The **nurse** always stands up for **their** patients.
Correct: **Nurses** always stand up for **their** patients.

Although writers could substitute *his or her* or *she or he* structures, these slow down the pace of writing. Some writers randomly alternate between genders for examples (as in this book). Still, whenever possible, recasting a sentence can obviate pronoun problems, which are not only confusing but potentially derogatory.

Even medical terms can be demeaning. Patients suffer from diseases; they are not "diabetics" or "the back surgery in Room 1284," as sometimes used for simplicity in case notes or hallway discussions. Writers should be vigilant not to reduce a human being to a disease or a bed. Words reflect reality—so make healthcare humanistic.

The last grammatical foible for this chapter is the *expletive*, which is a word that adds little to the meaning of sentences (except possibly emotion). Common examples are expletives used as interjections, such as "wow" or "gee." Few, if any, healthcare writers would ever interject "Hey, what a colon!"

A common form of expletive that does infiltrate many documents (healthcare and otherwise) actually derives from pronoun misuse. Structures with "there is/are/was/were" or "it is" (often starting sentences) can delay and potentially hide the sentence's intended subject and predicate. Some examples follow:

Expletive: **There** *are* many potential explanations for the embarrassingly high rate of infant mortality in our country.
Better: **Many factors** *potentially explain* the embarrassingly high rate of infant mortality in our country.

Expletive: **It** *is mandatory* for healthcare workers to wear protective gloves.
Better: **Healthcare workers** *must wear* protective gloves.

Notice that the reworked versions emphasize the true subjects and employ active verbs. Sometimes, expletive constructions better suit a topic, particularly if writers do not wish to emphasize the subject. Once again, as for voice and pronouns, one caveat remains: use expletives deliberately, not accidentally.

The next three chapters apply all the knowledge from these first eight chapters to examples of healthcare documents written for specific audiences. Keep alert for decisions on writing mechanics in these documents.

Chapter Summary

- Sentences are built from words, phrases, and clauses. Each of these grammatical structures carries its own level of emphasis. To facilitate comprehension, writers should carefully select which information belongs in which structure. Generally, emphasis increases in the following order: word, phrase, dependent clause, and independent clause.
- Independent (main) and dependent (subordinate) clauses determine the syntax of any sentence: simple, compound, complex, and compound-complex. Standard rules for joining clauses into sentences signal emphasis to the reader. Phrases do not affect a sentence's syntax.
- Relative clauses allow writers to signal to readers whether particular information is critical to understanding a sentence. The relative pronouns at the start of the clause, along with the presence or absence of commas, distinguish one type from the other.
- The underlying point of grammar and punctuation is to facilitate the transfer of meaningful information from writer to reader. Not applying the tools of writing mechanics can lead to confusion, particularly for audiences not conversant with the topic. Writers should pay extra attention to active vs. passive voice, dangling participles and other misplaced modifiers, confusing or sexist pronouns, and weak expletives that bury the message to be conveyed.

Exercises for Practice

1. Writers build sentences from three main building blocks: words, phrases, and clauses. Clauses are further delineated as independent (main) or dependent (subordinate). Moreover, some dependent clauses form a subset termed relative clauses. Identify the following types of syntactic building blocks:

   a. whoever administered the last dose
   b. tetanus
   c. the pharmacist identified a potential drug-drug interaction
   d. before being transferred to the step-down unit
   e. the patient's perforated bowel
   f. family care comprises services of a physician and a nurse practitioner
   g. which was not covered by insurance
   h. for outpatient therapy
   i. healthcare
   j. if the stitches should tear during routine movement

2. Correct or improve any of the following sentence structures, addressing any errors in grammar or punctuation (but some may be correct):

   a. It is important for you to read the label on this prescription-drug bottle before taking your first dose.
   b. The patient who was anemic required a blood transfusion.
   c. When the patient has been sedated the anesthetist begins monitoring her vital signs with his equipment.
   d. A patient's outlook, especially a positive one, can contribute to the success of their treatment.
   e. Without following the proper quality-control procedures, a patient could be exposed to a nosocomial infection, which is spread within a hospital setting and often difficult to combat.
   f. The dose was increased, however, the fever remained.
   g. The test, which the intern suggested, detected a possible dysfunction in the patient's liver.
   h. A battery of tests was run, none was conclusive.
   i. Although the accident victim quickly regained consciousness, his pulse rate remained dangerously low; therefore, the attending resident in the emergency room admitted him for overnight observation.
   j. In today's healthcare world, a diabetic may be seen by either a physician or a nurse practitioner.

The issue of injuries, particularly among teens, has received more attention, especially for student-athletes at any grade or level. Exercises 3, 4, and 5 relate to Figure 8.4, an excerpt from online information about injuries to the medial collateral ligament (MCL) within the knee.[4] (For easier use as an exercise, sentences in the excerpt have been numbered.)

Figure 8.4 Guidelines for Knee Injuries in Athletes

(1) Knee injuries often occur among active teens, especially athletes, and a torn medial collateral ligament (MCL)—a ligament that helps give the knee its stability—is a common knee injury.

(2) Teens who play contact sports, like football and soccer, are most likely to have a torn MCL. **(3)** The injury happens when the outside of the knee is struck, causing it to unnaturally bend inward (toward the other knee). **(4)** This creates tension on the MCL, a rope-like band, and it stretches or breaks in half.

(5) Someone with a partially or completely torn MCL might have swelling and pain within the first 24 hours of injury. **(6)** Fortunately, this injury can heal on its own with anywhere from 1 to 6 weeks of resting the joint.

(7) Most people with an MCL injury will still need to undergo rehabilitation ("rehab") therapy to help regain strength in the joint....

(8) The MCL is one of the four main ligaments in the knee joint. **(9)** It's located on the side of the knee that is closer to the other knee. **(10)** One end of the ligament is attached to the femur, while the other end is attached to the tibia.

(11) Together with the lateral collateral ligament (LCL), which is in the same location on the outside of the knee, the MCL helps prevent the overextension of the knee joint from side-to-side....

(12) Someone with a partially or completely torn MCL may or may not have symptoms, depending on the severity of the injury.

(13) Pain and swelling can be very intense initially, and some people (with more severe injuries) will have some instability when walking, feeling "wobbly" or unable to bear weight on the affected leg.

(14) Many people, especially those who are familiar with the injury or have torn a ligament before, report hearing a "pop" sound—the sound of the ligament tearing.

3. Identify the phrases, dependent clauses, and independent clauses. For any relative clauses, determine if they are restrictive or non-restrictive. How could some of these syntactic structures be modified to clarify the level of emphasis that the writer intends?

4. Determine the sentence types: simple, compound, complex, compound-complex. Does the distribution of sentence types work for the target audience and purpose? Identify and rewrite those sentences that might benefit from another structure.

5. Identify and categorize active vs. passive verbs, as well as expletive constructions. Revise any constructions that could be vague or hard to follow for the audience. Explain the reasoning for your choices.

References

1 Sibelious K. "Foreword." In US Centers for Disease Control and Prevention. *Voluntary Guidelines for Managing Food Allergies in Schools and Early Care and Education Programs.* Washington, DC: US Department of Health and Human Services; 2013. http://www.cdc.gov/HealthyYouth/foodallergies/pdf/13_243135_A_Food_Allergy_Web_508.pdf.

2 US Centers for Disease Control and Prevention. "Section 4. Putting Guidelines into Practice: Actions for Early Care and Education Administrators and Staff." In *Voluntary Guidelines for Managing Food Allergies in Schools and Early Care and Education Programs.* Washington, DC: US Department of Health and Human Services; 2013. http://www.cdc.gov/HealthyYouth/foodallergies/pdf/.

3 American Cancer Society. *What Is Targeted Therapy?* Atlanta: American Cancer Society. Last revised July 12, 2013. http://www.cancer.org/treatment/.

4 Atanda A, reviewer. *Medial Collateral Ligament (MCL) Injuries.* Wilmington, DE: TeensHealth, reviewed October 2012. http://kidshealth.org/teen/food_fitness/sports/.

9

GENERALIST AUDIENCE OF LAYPERSONS

Chapter Objectives

In this chapter, you will learn to:

- Analyze healthcare laypersons to describe a composite type
- Adjust strategies for constraints imposed by health literacy
- Recognize plain-writing techniques for healthcare laypersons
- Adapt healthcare educational materials to laypersons' needs
- Consider web-based systems that deliver timely information

Strategies for Laypersons

Too often, we healthcare writers focus so much on the complexities of research findings and clinical implications that we lose sight of the ultimate users—laypersons. Before we delve into specific techniques for laypersons (whether patients or family members) who lack scientific backgrounds, let's explore why effective communication with laypersons is critical today. With modern technologies, face-to-face encounters between patients and providers can easily be bypassed; moreover, real-time access bombards us with the latest information on drugs and surgeries. Is all this information helpful?

Managing this growing stockpile of information requires attention. Who decides what topics should be made available, and to which groups of people? Such information gatekeeping broaches a critical topic: acts of scientific inquiry and discovery bring ethical responsibilities. In other words, healthcare researchers (like other scientists) share the onus of "help[ing] the public to understand the knowable reach and limits, as well as the benefits and harms, of their work."[1] Consider a layperson's attempt to comprehend the package leaflet for a new prescription drug and you'll appreciate the ethical responsibility of healthcare communication. (Also see Chapter 13.)

Analysis of the Composite Layperson

A particular challenge when writing for laypersons (i.e., patients or families) is identifying a "common denominator" for their needs. In communication by mass media to a wide audience, rather than through tailored letters to individuals, healthcare writers face a daunting challenge: trying to reach the readers as individuals, not crowds. Often, healthcare writers (particularly in larger organizations) can avail themselves of separate information collected by marketing groups to characterize an audience subset. Otherwise, our recourse is to describe a wider layperson audience as a "composite" of characteristics most important to their understanding of the material.

Ideally, composites can be drawn from available demographic and psychographic details common in the advertising world. Demographic details include descriptors such as sex, age, and income, whereas psychographic details encompass lifestyle and social roles, personality traits, and personal values.[2] Pharmaceutical marketers, for instance, heavily invest into categorizing target audiences before devising advertisements. As another example, community hospitals glean a keen sense of the communities they serve, from the indigent marginalized at urban fringes to the affluent in upscale suburbia. These techniques can be adapted to various genres of healthcare writing.

Constraints of Health Literacy

One issue of paramount importance is health literacy. In general, writers for any audience must address the readers' degree of literacy, not just in terms of years of formal education but also as the ability to comprehend the text itself. This general literacy can be assessed (albeit sometimes crudely) by readability formulas weighing factors of syllables per word, average length of words, and complexity of syntax (as detailed in Chapter 12). In a healthcare context, literacy takes on the added dimensions of numeracy in tasks such as understanding laboratory assessments, reading food labels, administering the appropriate doses of over-the-counter drugs, and selecting from various insurance plans. Moreover, as technology advances, lack of familiarity with basic health concepts "can overwhelm even persons with advanced literacy skills."[3]

The US Department of Health and Human Services takes health literacy seriously. In fact, health communication (and its related information technology) is recognized to be "an integral part of the implementation and success of Healthy People 2020."[4] In terms of this ongoing initiative, health literacy is defined "as the degree to which individuals have the capacity to obtain, process, and understand basic health information needed to make appropriate health decisions."[5] Special populations—such as the elderly, immigrant, and indigent—pose unique challenges as some of the fastest growing groups. With low health literacy linked to reduced health outcomes,[6] healthcare writers must consider a spectrum of laypersons in determining composite audiences.

Plain-Writing Strategies for Laypersons

Earlier chapters have already discussed the myriad of writing techniques available for healthcare topics. From that foundation, this chapter aligns these techniques with the needs of a layperson audience. In general, underlying documents written for laypersons is one basic premise—plain language. This recognition of the need to present information more plainly originated with the US government stressing that information from agencies and departments must be conveyed meaningfully to those who use it. Spearheading this movement is the Plain Language Action and Information Network, a government-wide volunteer group working to improve communications from the federal government to the US public.[7] This group stresses that user-friendly documents should engage readers by careful consideration of audience type, clear writing to facilitate comprehension, and visual layouts that optimize messages. Modern technology offers potential synergy with plain writing toward improving health outcomes.[8]

For instance, a document written in plain language for a layperson audience might use second person (i.e., implied *you*) to communicate directly to readers as individuals, question-and-answer formats that anticipate readers' concerns, and step-by-step diagrams to clarify a sequential process. Writers should eschew compound-complex sentences and undefined technical terms, while also ensuring that all pronouns refer to an obvious noun (as detailed in Chapter 8). Moreover, section headings, bulleted or numbered lists, ample white (i.e., unused) space, and tables or graphs that complement text allow readers to focus on a document's key messages (as detailed in Chapter 7). Of course, writers would have selected in prewriting a strategy for organizing information in a manner suitable to the audience (as detailed in Chapters 3 and 4). Overall, software could be utilized to bring the information package to fruition.

Table 9.1 addresses 6 of the 11 attributes of effective healthcare writing identified within *Healthy People 2010*.[9] Four of these attributes—availability, reach, repetition, and timeliness—relate to distribution. A brochure on meningitis, for example, could be sent to homes of school-aged children or to parents attending a PTA meeting; alternatively, it may be linked through the school's website or published in a local newspaper. Decisions on distribution can affect a writer's job by influencing the document's setup: information to be displayed on a website needs to be more visual, whereas a brochure must not scare children with the seriousness of this disease.

Table 9.1 Selected Attributes of Effective Healthcare Writing

ATTRIBUTE	DEFINITION (ADAPTED FROM *HEALTHY PEOPLE 2010 REPORT*)
Distribution-focused:	
Availability	Placement in most accessible medium, whether print documents, electronic webs, or public ads
Reach	Availability of the information to the widest extent possible within the targeted audience group
Repetition	Continued delivery of consistent message across time for the same or subsequent user groups
Timeliness	Schedule of timed releases when audience is most needful of or receptive to that information
Writing-focused:	
Understandability	Levels of language and format (print or multimedia) appropriate for the targeted audience group
Cultural competence	Consideration of needs of special subpopulations, including ethnic, racial, and linguistic differences

The other two attributes—understandability and cultural competence—reinforce the earlier discussion of composite audiences, health literacy, and plain language. To be sure, understandability definitely underpins the very approach of audience analysis, and cultural competence is especially important in our increasingly multicultural society. Furthermore, words have both denotations (i.e., dictionary definitions) and connotations (i.e., emotional associations) that should be considered for all audiences, but especially for those subpopulations who might interpret words in different manners.[10]

As an example, a straightforward term like "side effect" denotes an untoward and detrimental outcome associated with a drug, such as the dry cough that can occur with the antihypertensive class called ACE (angiotensin-converting enzyme) inhibitors; however, many laypersons view side effects as conditions that will relentlessly strike (rather than occur in a small proportion of patients, as detailed in the leaflet). Alarmed patients might prematurely stop their medication, sometimes without informing healthcare providers. Additionally, unintentional stereotyping can detract from an otherwise effective message. Assuming illiteracy among a minority population, for instance, could result in language that insults rather than informs the target audience.

Document Examples for Laypersons

Layperson audiences receive healthcare information on many topics and through various means. In patient education, a particularly important focus, materials are typically distributed as leaflets or brochures available in providers' offices or community clinics. Although layperson-focused documents tend to be shorter, documents like educational materials sometimes require additional pages to allow adequate coverage of therapies and medical conditions.

Today's laypersons, however, continue to broaden in terms of their level of familiarity with healthcare information; many express an increasing desire for web-based information that is easily accessible. Web-based information itself needs to be presented concisely, but its non-linear logic (i.e., hyperlinked pages) requires a structure other than sequential.

The following two subsections of this chapter explore writing techniques to target laypersons as an audience group, using patient-education and web-based materials as key examples. For these examples, though, what sort of analysis of techniques should we conduct? Earlier chapters of this book already covered writing techniques appropriate to laypersons (and other audience groups): syntax (sentence structures), diction (word choice), person and voice, document strategy (e.g., comparison and contrast), signposting, and nesting of strategies. Techniques like these (as well as others) in the following examples provide an understanding of how healthcare writers can select and implement their own techniques for targeting a generalist audience.

Educational Materials for Laypersons

Many healthcare issues arise across the broad spectrum of laypersons, but almost everyone can relate to stress. A major step toward dealing with stress is understanding its physiological causes. Mental Health America, an organization sharing materials through linked websites of the US government, provides information on many topics for laypersons, including several on stress. Two connected versions of information related to stress (Figures 9.1 Ⓐ and 9.1 Ⓑ) can be accessed through the Mental Health America website.[11,12] Let's compare both versions for writing techniques.

Although writers should plan document approaches during the prewriting stage, analysis of documents often proceeds more easily backwards, as we'll now do for these two excerpts to determine strategies and techniques of the writer. Fortunately for writers, word-processing software comes equipped with tools that can check spelling, grammar, and readability. (Of course, writers should never rely blindly on

Figure 9.1 Ⓐ Advice to Laypersons for Managing Stress—Overview Version

Why Mental Health Matters

Some people think that only people with mental illnesses have to pay attention to their mental health.

But the truth is that your emotions, thoughts and attitudes affect your energy, productivity and overall health. Good mental health strengthens your ability to cope with everyday hassles and more serious crises and challenges. Good mental health is essential to creating the life you want.

Just as you brush your teeth or get a flu shot, you can take steps to promote your mental health. A great way to start is by learning to deal with stress.

How Stress Hurts

Stress can eat away at your well-being like acid eating away at your stomach. Actually, stress can contribute to stomach pains and lots of other problems, like:

- headaches
- insomnia
- overeating
- back pain
- high blood pressure
- irritability
- vulnerability to infection

Stress also can lead to serious mental health problems, like depression and anxiety disorders. If you think you have such a problem, you can get help.

Of course you can't magically zap all sources of stress. But you can learn to deal with them in a way that promotes the well-being you want—and deserve.

Learn more about how stress really hurts.

such tools, but instead consider them adjuncts to their own hands-on work; Chapter 12 details these techniques and their limitations.) In just a few seconds, we learn from this software that the overview excerpt (Figure 9.1 Ⓐ) contains 997 characters formed as 202 words, 12 sentences, and 16 paragraphs. Furthermore, we know that this excerpt organizes into about 5 characters/word, 14 words/sentence, and 2 sentences/paragraph. The detailed excerpt (Figure 9.1 Ⓑ) shows similar statistics.

What do these statistics mean? Clearly, the writer makes information available in *short* chunks—whether as short words, short sentences, or short paragraphs. Crisp and concise writing allows readers to digest information, being focused on content rather than distracted by structure. The structure of the writing should, in essence,

Figure 9.1 Ⓑ Advice to Laypersons for Managing Stress—Detailed Version

How Stress Hurts

Evolution was pretty savvy about danger. See a saber-tooth tiger, get moving! Today, flight—or fight, if necessary—still triggers major bodily changes, such as:

- Sugars in the bloodstream increase to supply energy
- Muscles tense so they're poised for action
- Heart beats faster to get blood pumping
- Digestion and other functions slow to save energy needed elsewhere

The problem is that our brains react to ominous loads of laundry and upcoming dentist visits like they were vicious predators. And the onslaught of today's stressors is fairly nonstop. When our bodies stay triggered for too long, lots of possible health problems can develop or worsen.

Stress may contribute to:

- high blood pressure
- heart disease and stroke
- decreased immune defenses
- cancer
- stomach problems
- poorer brain functioning

Stress also can lead to serious mental health problems, like depression and anxiety disorders.

Of course, you can't necessarily remove the sources of stress. But you can figure out ways to cope better with whatever comes your way. And decades of research suggest which steps are most likely to work.

be transparent to the reader. Moreover, software can also provide readability statistics: for this passage, the writing targets a reading level of 6th to 7th grade, a level typical for newspapers and other documents for laypersons.

While most words are indeed short, these words must also be reviewed for their deliberate selection—in other words, diction. The few overtly medical terms are familiar: "insomnia" and "depression" are examples. Use of commonplace allusions allows readers to associate stress with other common ailments: "Stress can eat away at your well-being like acid eating away at your stomach." Once readers begin to detach the stigma often associated with mental illness, they can be more responsive to specific recommendations for dealing with stressors in their lives.

Although the detailed excerpt displays similar readability statistics, subtle adjustments have been made due to the increased complexity of its content. Rather than assuaging the angst of readers who may be stressed already, as in the first excerpt, the second excerpt needs to explain through cause-and-effect reasoning. Most sentences in both excerpts are simple; the second, though, includes a few complex structures, such as this one: "When our bodies stay triggered for too long, lots of possible health problems can develop or worsen." In this instance, we reason forward—what is likely to happen if we don't address our stressors over time? A list provides the answer.

Besides cause-and-effect, another writing strategy in both excerpts is alterations in the selection of person for verb forms. Sentences that describe stress are often in third-person, such as the opening sentence of the first excerpt: "Some people think that only people with mental illnesses have to pay attention to their mental health." We can see a quick change to second person through most of both excerpts. Sentences like the following one allow readers to be the center of attention: "Good mental health is essential to creating the life you want." Note, too, the subtle shift to first person in the second excerpt, as in the following example: "The problem is that our brains react to ominous loads of laundry and upcoming dentist visits like they were vicious predators." Now readers can begin to connect with all of us who deal with stress. Who doesn't?

One technique potentially overdone, however, is starting simple sentences with coordinating conjunctions. Recall (from Chapter 8) that these seven words—*for, and, nor, but, or, yet, so*—allow two independent clauses to be joined (with a comma) into a compound structure. In layperson documents, some writers use coordinating conjunctions to start a simple sentence, as in this example: "And the onslaught of today's stressors is fairly nonstop." The underlying goal is to suggest relationships between independent clauses without inflating sentence length. Nonetheless, overuse of this modern technique can introduce an overly casual tone. Much like ending sentences with exclamation points, starting sentences with coordinating conjunctions is best limited.

Web-Based Information for Laypersons

In today's fast-paced world, laypersons expect (and even demand) that up-to-date healthcare information be available immediately. Fortunately, user-friendly portals to the Internet allow this demand to be met; in fact, the previous example illustrates how a web-based version could be used to entice readers into a longer text version with the space to provide more details on stress. Such educational materials for laypersons can be found through websites maintained by professional organizations, medical or nursing schools, and government agencies.

On websites, laypersons can choose to view selected materials on the screen or to print hard copies. Often, websites function as a depository of text documents that simply allow online access rather than paper mailings. The key distinction, though, is that web-based information (for laypersons and other groups) should facilitate easy access—audiences must find what they need. The "landing page" (first webpage encountered) is critical. Users must find the landing page designed with a welcoming appearance, but not overly cluttered with distractions, and clearly organized with navigational aids.

Websites with effective landing pages ease the entry of users into complicated systems, as is the case with the 2013-launched US Affordable Care Act (Figure 9.2). Despite supporters and critics of this legislation to extend healthcare, the website's landing page[13] visually succeeds in enticing potential users to explore the site

Figure 9.2 Landing Page for US Affordable Care Act

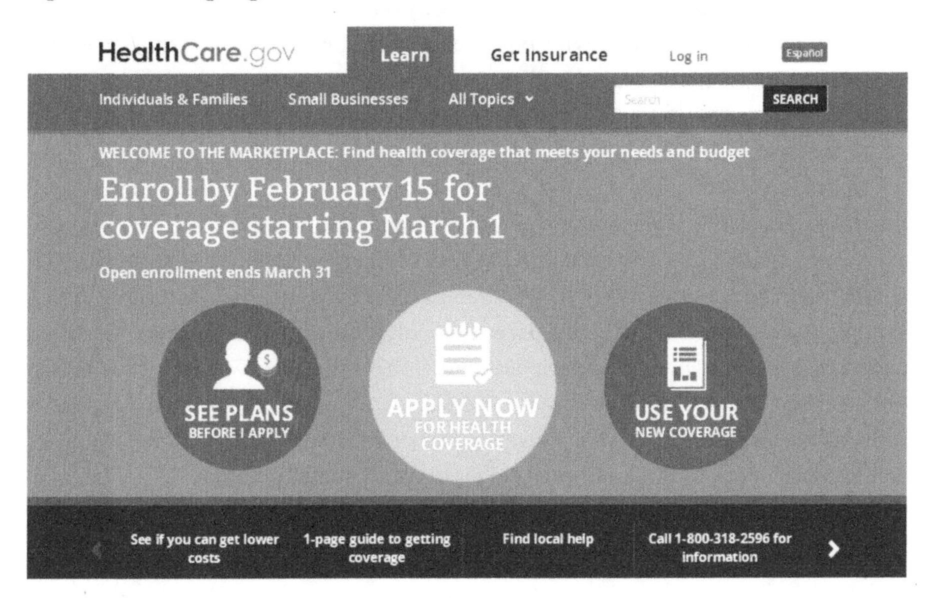

(notwithstanding technical glitches in its initial launch). Three large circles, each with its own simplified icon, show the pathways to find information on plans, to enroll once a plan has been selected, and to use the coverage afforded by that selected plan. The horizontal menu across the top of the page lets visitors find information grouped for either individuals and families or small businesses. Rather than being cluttered with small text, this landing page emphasizes key points, such as enrollment deadlines, through well-positioned words.

Navigational features, moreover, are crucial in volatile or serious situations, such as the recall of a prescription drug due to serious, unexpected side effects. Consider the case of Vioxx™ (rofecoxib), a member of the NSAID (nonsteroidal anti-inflammatory drug) class intended to alleviate painful conditions like arthritis. After completion of required safety assessments, this marketed drug was associated in the wider population with an increased risk for cardiovascular events, including heart attack and stroke. (By then, it had been on the market for over five years.) Its manufacturer voluntarily withdrew the drug from US and international markets where it had been approved for sale. Subsequently, safety concerns extended to the full NSAID class; the US Food and Drug Administration (FDA) issued guidance on communicating these risks to consumers through having the manufacturer revise the package labeling.[14]

On its website, the FDA updated special links for information leading up to and following the withdrawal of Vioxx. Individuals could find the latest information by date of posting; colored icons of "NEW!!" alerted laypersons to updated documents. Complex information surrounding the withdrawal of Vioxx was conveyed to concerned, frightened laypersons through an easily navigable structure (Figure 9.3). Its question-and-answer format with bold text directed readers to the most relevant questions.[15] Importantly, the FDA did not shy away from answering difficult questions on implications for consumers who had already been taking this medication under their providers' care.

In this tense scenario, the FDA successfully used writing techniques to ensure that laypersons could access critical information. Use of first person ("We encourage people taking Vioxx to contact their physician") engendered a reassuring level of commitment and responsibility. The readability of this passage is a tenth-grade level. Complex terms are clarified with parenthetical definitions, as in "gastrointestinal (stomach) bleeding" and "long period of time (longer than two weeks)." The average sentence length of 15 words complements such simplification. Some formal remnants reflect the balancing act to be achieved between legal/regulatory statutes and user comprehension, as in this additional example not included in Figure 9.3:

Figure 9.3 Web-Based Document on the Safety of Vioxx (Excerpt)

Vioxx (rofecoxib) Questions and Answers

1. What action did Merck take today?*

Merck announced a voluntary worldwide withdrawal of Vioxx (rofecoxib).

2. What is Vioxx?

Vioxx is a COX-2 selective nonsteroidal anti-inflammatory drug (NSAID). Vioxx is also related to the nonselective NSAIDs, such as ibuprofen and naproxen. Vioxx is a prescription medicine used to relieve signs and symptoms of arthritis, acute pain in adults, and painful menstrual cycles. [...]

5. What should I do if I am currently taking Vioxx?

The risk that an individual patient will suffer a heart attack or stroke related to Vioxx is very small. We encourage people taking Vioxx to contact their physician to discuss discontinuing use of Vioxx and alternative treatments. Any decision about which drug product to take to treat your symptoms should be made in consultation with your physician based on an assessment of your specific treatment needs.

6. What are the likely long-term health effects, if any, of taking this product?

The new study shows that Vioxx may cause an increased risk in cardiovascular events such as heart attack and strokes during chronic use. [...]

12. Does today's action suggest that other drugs in the same class are dangerous?

The results of clinical studies with one drug in a given class do not necessarily apply to other drugs in the same class. All of the NSAIDs have risks when taken chronically, especially of gastrointestinal (stomach) bleeding, but also liver and kidney toxicity. Patients using these drugs for a long period of time (longer than two weeks) should be under the care of a physician. [...]

14. Can my pharmacist continue to fill my prescription for Vioxx?

No, Merck is initiating a market withdrawal in the United States to the pharmacy level. This means Vioxx will no longer be available at pharmacies.

"Today" refers to September 30, 2004, the date of the voluntary withdrawal of Vioxx.

Question 3. Did FDA require this action?

No, Merck made this decision independent of input from FDA. *The Agency has not had an opportunity to review the data from the study that was stopped in the depth that Merck has, but agrees with the company that there appear to be significant safety concerns for patients, particularly those taking the drug chronically.* FDA plans to work closely with Merck to coordinate the withdrawal of this product from the US market. (*emphasis added*)

Writers familiar with guidelines of the FDA and other regulatory bodies can relate to wording like this identified sentence. However, since the situation required rapid reply, the FDA needed to avoid drawing conclusions without sufficient evidence. Clauses like "there appear to be significant safety concerns" allow safety issues to be raised without definitive causal linkage. Moreover, this wording addresses interests of various secondary audiences (e.g., lawyers, competitors) while remaining focused on the primary audience (i.e., patients and their families). Those layperson viewers were able to find information that they needed quickly.

Web-based information targeting laypersons, though, needs to consider those with limited health literacy (as discussed earlier in this chapter) and those uncomfortable with the Internet. These users may be overwhelmed or frustrated. Recommendations for designing websites in these situations include organizing information with a clear navigational plan, keeping complex content in relatively simple displays, and focusing on decision-making and other actions of the users.[16]

Although especially pertinent to the layperson audience, these tips should also be considered for other audience groups. For example, administrators need easy access to information because of their own job responsibilities. While health literacy would not be expected to be of concern, any website should also provide easy navigation for these busy managers (as covered in the next chapter).

Chapter Summary

- Layperson audiences vary, so writing for them can be challenging. Sometimes, writers can use demographic and psychographic details to construct a composite audience. Even so, writers must remember that readers are individuals.

- Literacy—whether generally for reading or specifically in healthcare—plays an important role in documents that convey information to layperson audiences. The implications of health literacy can be especially critical for special populations like the elderly or those from non-traditional cultures.

- Plain writing represents a movement to simplify documents from the government to the public. Techniques for achieving this goal include wording that facilitates comprehension of complicated material, as well as optimizing visual layout in ways that reinforce or enhance that information.

- Patient-education materials represent a common form of documents written for generalist audiences. Words, sentences, and paragraphs should be shorter so that readers focus on the information, not sentence structure. Signposting of writing strategies assists the reader in following the information.

- In today's technological age, many laypersons have come to expect healthcare information to be available immediately—as on the Internet. Websites need to be designed with a navigational logic that allows readers to find whatever materials they want as they first encounter the landing page.

Exercises for Practice

1. Immunology is a complicated topic (particularly for laypersons), given different cell types that function in a cascade, along with internal and external components that can trigger this cascade. While most people may be aware of allergies, they may not understand autoimmune diseases—both deleterious expressions of the immunological system. Here are two descriptions of autoimmune disease, taken from websites for laypersons. Critique their effectiveness if the description were designed to inform individuals recently diagnosed with an autoimmune disease. Substantiate your decision with strategies and other writing techniques.

> **Description 1:** *Our bodies have an immune system, which is a complex network of special cells and organs that defends the body from germs and other foreign invaders. At the core of the immune system is the ability to tell the difference between self and nonself: what's you and what's foreign. A flaw can make the body unable to tell the difference between self and nonself. When this happens, the body makes autoantibodies (AW-toh-AN-teye-bah-deez) that attack normal cells by mistake. At the same time special cells called regulatory T cells fail to do their job of keeping the immune system in line. The result is a misguided attack on your own body. This causes the damage we know as autoimmune disease.*[17]

> **Description 2:** *Your body's immune system protects you from disease and infection. But if you have an autoimmune disease, your immune system attacks healthy cells in your body by mistake. Autoimmune diseases can affect many parts of the body.... No one is sure what causes autoimmune diseases. They do tend to run in families.... The classic sign of an autoimmune disease is inflammation, which can cause redness, heat, pain and swelling.... Treatment depends on the disease, but in most cases one important goal is to reduce inflammation.*[18]

2. In the past, patients taking prescription drugs had to strain their eyes to read the miniscule print of folded package inserts that were written for experts. Now, the FDA requires that pharmaceutical companies and pharmacies provide a simpler version for laypersons. Furthermore, the FDA provides online guides for drugs with potential safety issues. The excerpt in Figure 9.4 is a guide excerpt for Humira® (adalimumab, AbbVie Inc.), an injectable drug with immunological actions.[19] Identify techniques that were used to target laypersons. Would you suggest revisions? If so, specify techniques and justify your reasons.

Figure 9.4 Layperson-Focused Prescribing Information on Humira

MEDICATION GUIDE
HUMIRA® (Hu-MARE-ah)
(adalimumab)
Injection

Read the Medication Guide that comes with HUMIRA before you start taking it and each time you get a refill. There may be new information. This Medication Guide does not take the place of talking with your doctor about your medical condition or treatment.

What is the most important information I should know about HUMIRA?

HUMIRA is a medicine that affects your immune system. HUMIRA can lower the ability of your immune system to fight infections. **Serious infections have happened in people taking HUMIRA. These serious infections include tuberculosis (TB) and infections caused by viruses, fungi or bacteria that have spread throughout the body. Some people have died from these infections.**

- Your doctor should test you for TB before starting HUMIRA.
- Your doctor should check you closely for signs and symptoms of TB during treatment with HUMIRA.

You should not start taking HUMIRA if you have any kind of infection unless your doctor says it is okay.

Humira® is a registered trademark of AbbVie Inc., North Chicago.

3. Websites provide access to information on many levels—provided that navigation is logical and clear. As discussed in this chapter, *Healthy People* seeks to assess and improve various health indicators. Major updates occur on decennial years: 2000, 2010, 2020, and so on. Figures 9.5 Ⓐ and 9.5 Ⓑ provide the "landing pages" (first webpages encountered) for the 2010 version[20] and the 2020 version[21] of the *Healthy People* initiative. Compare and contrast changes from the earlier to the more recent version. Have the intended audiences broadened?

Figure 9.5 Ⓐ Navigational Plan for *Healthy People 2010* Website

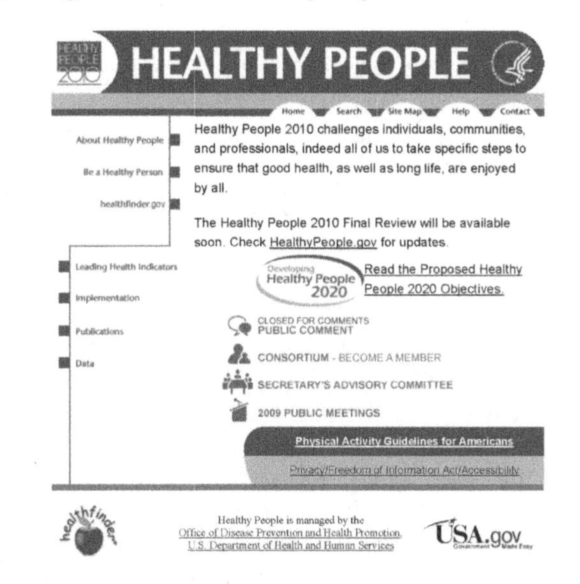

Ⓑ Revised Navigational Plan for *Healthy People 2020* Website

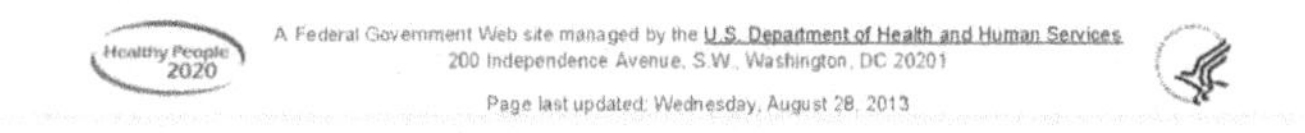

References

1 Reiser SJ. The ethical challenges of explaining science. *American Medical Writers Association Journal*. 2003;18(3): 96-8. http://www.amwa.org/.

2 Stovall JG. Writing advertising copy. In: *Writing for the Mass Media*. 7th ed. Boston: Pearson Education; 2009; chap 10.

3 Office of Disease Prevention and Health Promotion. *Health Communication, Health Literacy, and e-Health*. Washington, DC: US Department of Health and Human Services; last updated January 1, 2014. http://www.health.gov/communication/literacy.

4 Office of Disease Prevention and Health Promotion. *Healthy People 2020*. Washington, DC: US Department of Health and Human Services; November 2010. ODPHP Publication No. B0132. http://www.healthypeople.gov/2020.

5 Committee on Health Literacy. *Health Literacy: A Prescription to End Confusion*. Washington, DC: Institute of Medicine; 2004. Cited in: Kunter M, Greenberg E, Jin Y, Paulsen C, White S. *The Health Literacy of America's Adults: Results from the 2003 National Assessment of Adult Literacy*. Washington, DC: US Department of Health and Human Services; September 2006. NCES 2006-483. http://www.nces.ed.gov.

6 Beckman ND, Sheridan SL, Donahue KE, Crotty K. Low health literacy and health outcomes: an updated systematic review. *Annals of Internal Medicine*. 2011 Jul 19; 155(2):97-107. Available at: http://www.annals.org.

7 Plain Language Action and Information Network. *Federal Plain Language Guidelines*. Washington, DC: Plain Language Action and Information Network; last updated May 2011. http://www.plainlanguage.gov/.

8 Office of Disease Prevention and Health Promotion. *Health Communication and Health Information Technology*. Washington, DC: US Department of Health and Human Services; last updated November 13, 2013. http://www.healthypeople.gov/2020.

9 Office of Disease Prevention and Health Promotion. *Healthy People 2010: Objectives for Improving Health*. Washington, DC: US Department of Health and Human Services; November 2000. http://www.healthypeople.gov/2010.

10 Bonk RJ. Societal framework of health and medicine: recommendations for medical writers. *American Medical Writers Association Journal*. 2004;19(1):12-5.

11 Mental Health America. *Live Your Life Well*. Alexandria, VA: Mental Health America; last updated May 2014. http://www.mentalhealthamerica.net.

12 Mental Health America. *How Stress Hurts*. Alexandria, VA: Mental Health America; last updated May 2014. http://www.mentalhealthamerica.net.

13 Health Insurance Marketplace. *Welcome to the Marketplace*. Baltimore, MD: US Centers for Medicare & Medicaid Services; February 2014. http://www.healthcare.gov.

14 US Food and Drug Administration. *COX-2 Selective (Includes Bextra, Celebrex, and Vioxx) and Non-Selective Non-Steroidal Anti-Inflammatory Drugs (NSAIDs)*. Silver Spring, MD: US Department of Health and Human Services; last updated November 27, 2012. http://www.fda.gov/drugs/drugsafety/.

15 US Food and Drug Administration. *Vioxx (rofecoxib) Questions and Answers*. Silver Spring, MD: US Department of Health and Human Services; last updated December 30, 2009. http://www.fda.gov/drugs/drugsafety/.

16 Office of Disease Prevention and Health Promotion. *Health Literacy Online: A Guide to Writing and Designing Easy-to-Use Health Web Sites*. Washington, DC: US Department of Health and Human Services; 2010. http://www.health.gov/healthliteracyonline/.

17 Office on Women's Health. *Autoimmune Diseases Fact Sheet*. Washington, DC: US Department of Health and Human Services; last updated July 16, 2012. http://www.womenshealth.gov/.

18 US National Library of Medicine. *Autoimmune Diseases*. Bethesda, MD: National Institutes of Health; last updated January 27, 2014. http://www.nlm.nih.gov/medlineplus/autoimmunediseases.html.

19 US Food and Drug Administration. *Medication Guide: Humira® (Hu-MARE-ah) (adalimumab) Injection*. Silver Spring, MD: US Department of Health and Human Services; last updated May 12, 2014. http://www.fda.gov/drugs/drugsafety/.

20 Office of Disease Prevention and Health Promotion. *Healthy People 2010*. Washington, DC: US Department of Health and Human Services; November 2000. http://www.healthypeople.gov/2010.

21 Office of Disease Prevention and Health Promotion. *Healthy People 2020*. Washington, DC: US Department of Health and Human Services; last updated August 28, 2013. Available at: http://www.healthypeople.gov/2020.

10

MANAGERIAL FOCUS OF ADMINISTRATORS

Chapter Objectives

In this chapter, you will learn to:

- Highlight the administrative elements of healthcare situations
- Select writing techniques suitable to the administrator's focus
- Follow requirements for persuasive proposals that seek funding
- Clarify complicated processes administered by office managers
- Repackage technical details into concise executive summaries

The Growing Role of Healthcare Administrators

In many respects, healthcare can be considered a basic human right. Even though health cannot be guaranteed, shouldn't everyone have access to care? Nonetheless, care comes at a cost: budgets for individual persons and national governments alike must allocate limited resources across seemingly unlimited needs. Laypersons and practitioners both need to consider insurance coverage, copay percentages, and treatment formularies (to name a few) when selecting a new prescription medication, for instance. Efficacy and safety no longer suffice, now that the decision triad includes cost. Even researchers must consider monetary realities: exploratory treatments without approved indications may be sliced from the budget. That's the very nature of economics, after all.

Table 10.1 Writing Scenarios for Healthcare Administrators

ADMINISTRATOR ROLE	DOCUMENT EXAMPLE
Marketing agent for pharmaceutical firm	Developmental goals of a new drug vs. current competition
Claims processor in insurance company	Hard-copy or online compendium that details options within plans
Policy analyst within government office	Budgetary implications regarding expansion of federal assistance

This focus led to the burgeoning of the next healthcare audience—administrators. Applying expertise in handling resources and timelines, administrators allow researchers to focus on research and practitioners to focus on practice. Consider a drug company: the corporate administrators in many managerial and support roles facilitate the efforts by scientists in research and development. This allocation of human resources maximizes the stockholders' return on investment, which is critical for corporate survival. Analogously, administrators handling piles of paperwork in the offices of healthcare providers allow the practitioners to focus on their patients. Moreover, administrators at insurance firms, as another example, shepherd requests for coverage and reimbursement, particularly of note for health maintenance organizations (HMOs).

Given the ever-increasing role of administrators, healthcare writers need to craft documents to facilitate their managerial work. What do administrators need and want? This chapter explores answers to this question.

Establishing the Administrator Focus

Think about a parent contacting his insurance group to determine why his child's antibiotic prescription was denied for coverage. At this stage, the conversation would not concern the growing problem of resistance due to incorrect prescription of antibiotics for viral (not bacterial) infections. Potentially, the pharmacy might have intervened in such a situation. The anxious, angry father is not focused on public health—he wants to protect the health of his ill child. The claims processor

WRITING STRATEGIES	WRITING MECHANICS
Overall chronological strategy with order of importance for development goals; appendix of pharmacological properties needed to fulfill these goals; graphic to present timeline	Clear language emphasizing business aspects; complex sentences to reflect decision points in drug's development; active voice to identify those responsible for critical steps
Overall compare-contrast by wholes (i.e., plans) before parts (i.e., options); logos superseding pathos; figure of follow-up steps for handling unsatisfied clients	Imperative (i.e., command) mood aligned with bulleted items instead of regular text; simple sentences for action steps, but complex syntax for dependencies (i.e., if-then)
Problem-method-solution to clarify existing situation, its analysis, and policy change; nested classification-partition for easier handling of details; ethos for balanced analysis	Simple sentences to balance complicated details, although compound-complex needed to handle certain details; use of restrictive clauses to clarify expenditures by plan items

would need to be able to explain reasons behind the coverage refusal, to reconsider if the denial was not consistent with coverage policy, and to expedite notification of options to the pharmacy. In addition, the processor would need to collect policy numbers, plan levels, and so on.

Now consider the processor's position. Her focus would be on computer screens, coverage specifics, and decision trees. Perhaps she would also need training materials for dealing with irate clients. Those contextual needs of the administrator need translation into appropriate writing techniques to address, within policy restrictions, the complaint from a layperson. Certainly, at this hypothetical stage, we can only suggest approaches for documents needed by the administrator; prewriting would be the proper time to tailor strategies and mechanics for this volatile situation. The writer's goal, nonetheless, would be to provide quick access to procedural (rather than technical) information. Reasonably, the writer could assume that the administrator has the requisite knowledge for her role; double-checking such assumptions, though, would be wise.

Another complication is that healthcare administrators vary in the specific roles that they may fill in their organizations. Table 10.1 presents several scenarios regarding administrators in various healthcare arenas. As shown for these scenarios, writers must assess the particular role of the administrators, given their wide range of responsibilities; specifying purpose early facilitates identification of appropriate strategies and mechanics. Then, the writer could select an overall structure for the document. In Table 10.1, the first and third scenarios use nested strategies, whereas the second repeats the compare-contrast strategy in two fashions (whole-by-whole and part-by-part) to meet two different uses of the information. As for mechanics, these three scenarios require clear language to avoid confusion under pressure. Emphasizing simple sentences might be one approach, but the other syntactic structures might also be appropriate. Bullets and other format techniques would also come into play.

Document Examples for Administrators

Let's now take a look at real (not hypothetical) documents that target the unique needs of healthcare administrators. As seen in Table 10.1, administrators play a wide range of healthcare roles. Since documents targeting these administrators are similarly varied, the following three examples represent only a small number of possibilities. Regardless, each exemplifies typical strategies, writing mechanics, and design considerations pertinent to documents targeting an administrator audience.

Proposal Requirements for Financial Grants

One of the most crucial—albeit complicated—documents for administrators is a proposal to the government or a philanthropic group for a grant. External funding is often necessary for smaller agencies, such as community health centers, with limited resources. Grant applications need to be persuasive, but much of that persuasion comes from logos rather than pathos. To judge the best stewardship for support, funding agencies consider hard data: number of individuals at risk; operational budgets for previous, current, and future programs; and plans to assess outcomes. With many applicants for limited funds, agencies must be able to award grants in an unbiased, ethos-driven manner. Anecdotes and other pathos-driven arguments have limited applications in grant proposals, relative to requirements for logos-driven information.

Many grants are funded through healthcare agencies of the US government, such as the Department of Health and Human Services. Given the variety of grants available, sub-agencies often channel funding for particular healthcare needs. Mental health, as one example, has particular needs for helping individuals at risk for these sometimes ignored conditions. In this case, the Substance Abuse and Mental Health Services Administration (SAMHSA) provides a central point for information and support. Figure 10.1 Ⓐ provides an overview of the application process for grants in this area.[1] Notice the use of hierarchical headings, short sentences, second person, and bulleted lists. Links to related websites are included; one accesses a training video on the process itself.

Figure 10.1 Ⓐ Guidelines for Applying for Government Grants

These resources will help you find and apply for funding opportunities.

Eligibility

Only domestic public and private non-profit entities can apply for SAMHSA grants. Check for any additional eligibility requirements in the specific SAMHSA Request for Applications (RFA), or on Grants.gov, to see if you are eligible to apply for that grant.

Choose a Funding Opportunity

Search SAMHSAs RFAs on the Grant Announcements page or on Grants.gov.

To find a SAMHSA funding opportunity on Grants.gov, enter either the Funding Opportunity Number (SAMHSA's RFA number) or the Catalog of Federal Domestic Assistance (CFDA) number. Both numbers are on the cover page of the RFA.

Each RFA includes specific instructions for completing your application.

Information on Electronic Submission

Visit the Submitting Your Application page for information about the three separate registration processes you must complete to submit a grant application to SAMHSA.

Note:

SAMHSA requires that applications be submitted electronically through Grants.gov.

SAMHSA will not accept or consider any applications sent by email or fax, or hand-carried.

Watch a Grants.gov Webcast on How to Register, Find, and Apply

Grants.gov Applicant Training Video—Register, Find, and Apply will help you learn more about registering, searching, and applying for federal grant opportunities.

A specific grant will have its own particular requirements, detailed in a "Request for Application" (RFA). The 86-page RFA for funds under Project LAUNCH—Linking Actions for Unmet Needs in Children's Health—lists specific documents to be included, as shown in Figure 10.1 Ⓑ.[2] Despite the high level of detail in this instructional section, the information is provided clearly through writing techniques such as bulleted lists, bolded words, short paragraphs, and second person. Not shown in this excerpt are instructions, sometimes with sample documents, for materials to be appended to the grant application. Although assembling the documents may be time-intensive, this roadmap facilitates the process for an administrator overseeing the grant application.

Figure 10.1 Ⓑ Documentation Requirements for Project LAUNCH Grants

- **Abstract** - Your total abstract must not be longer than 35 lines. It should include the project name, population(s) to be served (demographics and clinical characteristics), strategies/interventions, project goals and measurable objectives, including the number of people to be served annually and throughout the lifetime of the project, etc. In the first five lines or less of your abstract, write a summary of your project that can be used, if your project is funded, in publications, reports to Congress, or press releases.
- **Table of Contents** - Include page numbers for each of the major sections of your application and for each attachment.
- **Budget Information Form** - Use SF-424A. Fill out Sections B, C, and E of the SF-424A. A sample budget and justification is included in Appendix H of this document.
- **Project Narrative and Supporting Documentation** - The Project Narrative describes your project. It consists of Sections A through G. Sections A through G together may not be longer than 30 pages. (Remember that if your Project Narrative starts on page 5 and ends on page 35, it is 31 pages long, not 30 pages.) More detailed instructions for completing each section of the Project Narrative are provided in "Section V - Application Review Information" of this document.

  The Supporting Documentation provides additional information necessary for the review of your application. This supporting documentation should be provided immediately following your Project Narrative in Sections H - J. There are no page limits for these sections, except for Section I, Biographical Sketches/Job Descriptions. Additional instructions for completing these sections are included in Section V under "Supporting Documentation." Supporting documentation should be submitted in black and white (no color).
- **Attachments 1 through 9** - Use only the attachments listed below. If your application includes any attachments not required in this document, they will be disregarded. Do not use more than a total of 30 pages for Attachments 1, 3 and 4 combined. There are no page limitations for Attachments 2 and 9. Do not use attachments to extend or replace any of the sections of the Project Narrative. Reviewers will not consider them if you do. Please label the attachments as: Attachment 1, Attachment 2, etc.

Processing Coverage and Reimbursement Claims

Often, administrators in the office of healthcare providers handle procedures that relate to coverage and/or reimbursement for therapies. In this way, the administrators can capitalize on their own expertise while allowing practitioners to do the same. An example comes from a Medicare document for handling exceptions to caps (i.e., coverage limits) on therapies charged to this government plan primarily for senior citizens.[3] A numbering system allows all participating billing offices to refer to the same procedural guidelines easily. This excerpt (Figure 10.2) was selected from an online menu of the US Centers for Medicare & Medicaid Services (CMS); only the first page is presented.

Figure 10.2 Medicare Process for Exception to Therapy Caps

Medicare Learning Network
MLN Matters
CMS
Information for Medicare Fee-for-Service Health Care Professionals

MLN Matters Number: MM4364	**Related Change Request (CR) #: 4364**
Related CR Release Date: February 15, 2006	**Effective Date: January 1, 2006**
Related CR Transmittal #: R52BP, R140PI, R855CP	**Implementation Date: No later than March 13, 2006**

Note: This article was updated on November 1, 2012, to reflect current Web addresses. All other information remains unchanged.

Therapy Caps Exception Process

Important Note: Recent legislation extended the therapy cap exceptions for calendar year 2007. For details on the 2007 exceptions and process, see the MLN Matters article MM5478 at http://www.cms.gov/outreach-and-education/medicare-learning-network-mln/mlnmattersarticles/downloads/MM5478.pdf on the CMS site.

Provider Types Affected

Providers, physicians, and non-physician practitioners (NPPs) who bill Medicare contractors (fiscal intermediaries (FIs) including regional home health intermediaries (RHHIs), and carriers) under the Part B benefit for therapy services

Key Points

- Effective January 1, 2006, a financial limitation (therapy cap) was placed on outpatient rehabilitation services received by Medicare beneficiaries. These limits apply to outpatient Part B therapy services from all settings except the outpatient hospital (place of service code 22 on carrier claims) and the hospital emergency room (place of service code 23 on carrier claims). Outpatient rehabilitation services include:
    - **Physical therapy** - including outpatient speech-language pathology: Combined annual limit for 2006 is $1,740; and
    - **Occupational therapy** - annual limit for 2006 is $1,740.
- In 2006 Congress passed the Deficit Reduction Act (DRA), which allows the Centers for Medicare & Medicaid Services (CMS) to grant, at the request of the individual enrolled under the Part B benefit or a person acting on behalf of that individual, **exceptions to therapy caps for services provided during calendar year 2006**, if these services meet certain qualifications as medically necessary services (Section 1833(g) (5) of the Social Security Act).

Despite this document being procedural in nature, its layout and design allow the details to be followed without strain. Identifying numbers and effective dates appear atop the first page, followed by a text box that highlights the most recent update. Reasoning for the exception process is briefly provided, followed by a list of affected providers by type of service. Then, bulleted points provide procedural details in a hierarchical manner (which extends throughout the full six pages of this document). Bolding of text focuses the reader on particular issues; generous use of white space keeps the procedure from seeming too dense. A disclaimer, typically required with guidelines, appears as a footnote (to each page); despite its smaller font, this footnote meets the additional needs of secondary audiences who might also read this procedure.

Overall, this excerpt demonstrates how document layout and design can facilitate the transfer of complicated details like a billing procedure to the administrators who must follow these processes. This document, though, could go a step further by supplementing the procedure with a flow diagram. Although cross-references to other relevant manuals allow administrators to find related information, keeping the users at one website may be preferable. Of course, writers (like the one for this document) need to balance the length of individual documents with the typical users' needs. Regardless of the need for some slight improvements, this document does a solid job of clearly presenting a process.

Summarizing Details for Executives

The third example comes from a formal report (main text of 37 pages) issued by the World Health Organization (WHO), an organizational body operating under the aegis of the United Nations. Understandably, WHO operates many initiatives cross the globe; maintaining oversight from an administrator's perspective can be challenging, especially given WHO's fiduciary responsibility. Therefore, a Task Force on Resource Mobilization and Management Strategies was charged to review the organization's financial flexibility for addressing health concerns that can change daily.

While assembling the full report was undoubtedly a time-intensive task for the committee, comprehending the full report may have been an even more time-intensive endeavor for the administrators at WHO. To facilitate their work, an executive summary that digested the findings and recommendations was included in the front matter of the overall report.[4] Figure 10.3 provides excerpts from the original summary of two pages that illustrate writing techniques for executive audiences.

Figure 10.3 Executive Summary from World Health Organization Report

The Task Force on Resource Mobilization and Management Strategies conducted extensive analyses and reviewed options for action to overcome the challenges that were identified. In summary, the analyses revealed the following....

WHO has been successful in generating resources, receiving US$1.1 billion in assessed and core voluntary contributions and nearly US$ 3 billion in other voluntary resources every biennium. Financing is nevertheless uneven over technical areas and locations. Further, resource predictability and transparency is limited. Although every part of WHO is involved in resource mobilization, it is poorly coordinated, and the corporate perspective is missing. The absence of common tools and processes for quality assurance and for tracking individual resource mobilization efforts might have an adverse effect on financing risk management and could eventually strain donor relations....

On the basis of this analysis, the Task Force has made recommendations in 10 broad areas:

1. to make the PB the main tool for resource mobilization, performance and resource management;
2. to maximize the potential of the Financing Dialogue for resource mobilization and alignment;
3. to improve, coordinate and harmonize tools and processes for quality assurance and for tracking the "pipeline";
4. to professionalize WHO's resource mobilization;
5. to reduce vulnerability by broadening the donor base, including involving the humanitarian forum;
6. to improve resource mobilization and management at country level;
7. to devise explicit, transparent mechanisms for resource allocation and management;
8. to improve WHO programme and project delivery and performance to facilitate resource mobilization and management;
9. to improve the monitoring and evaluation of resource flows and performance; and
10. to better position WHO and to communicate its uniqueness, contributions and deliverables.

The specific recommendations in each of these areas are given in the report and its annexes.

The executive summary opens with a brief outline of the charge and procedures of the Task Force. Highlights of the findings then follow; these findings are recast in terms of "resource predictability" and "risk management" as most pertinent to administrators. Rather than listing financial specifications for particular health programs, this summary positions itself above details, thereby allowing conclusions and recommendations to be the foci for the audience. After summarizing similar information, the executive summary lists the main recommendations of the Task Force (details of which can be found in the text of the full report).

For healthcare practitioners implementing medical interventions on behalf of the United Nations, this executive summary feels insufficient. It is. Executives overseeing the administration of WHO's resources do not need that level of programmatic detail, which would detract from their financial focus. In fact, additional administrative details at the country level have been relegated to an appendix at the end of the report. Nevertheless, the executive summary fulfills the needs of the administrator audience. Techniques for meeting the needs of the practitioner audience are covered in the next chapter.

Chapter Summary

- Administrators, who focus on managerial and related details rather than medical and scientific data, play important roles in managing today's healthcare resources. Procedures, finances, and resources form their main interest areas; handling these managerial details allows practitioners to focus on healthcare itself.

- Given their managerial focus, administrators need documents that can concisely translate medical information into resource considerations. Crisp language, visual cues, and bulleted lists are just some of the appropriate writing techniques to be considered for administrators.

- Many healthcare agencies need external funding to support their initiatives, so administrators in those agencies must know how to prepare grant applications that will persuade the awarding group. While persuasive in nature, proposals rely on hard evidence over anecdotal support.

- Procedures for handling healthcare costs, expenditures, and reimbursements can be daunting to those unfamiliar with such intricacies. Effective layout and design can help to organize the complicated procedures beyond considerations of syntax and other writing mechanics.

- Administrators rely on translations of medical details into managerial concerns such as resource allocation. Hence, formal reports typically begin with an executive summary that provides that managerial focus.

Exercises for Practice

1. Administrators need to remain versed in continual changes in healthcare reform. Government legislation, insurance plans, prior authorizations, and reimbursement processes are just a few examples. For this reason, many administrators attend professional conferences. Critique information posted about the 2014 annual meeting of the Drug Information Association (Figure 10.4)[5] and offer suggestions for improvement.

Figure 10.4 Description of Annual Meeting of the Drug Information Association

About DIA 2014

The DIA 2014 50th Annual Meeting is the largest multidisciplinary event that brings together a community of life sciences professionals at all levels and across all disciplines involved in the discovery, development, and life cycle management of medical products all with a common goal to foster innovation that will lead to the development of safe and effective medical products and therapies to patients.

DIA 2014 50th Annual Meeting will:

- Provide a global unbiased platform to raise the level of discussion to facilitate innovation that will change the future of health care
- Provide comprehensive coverage of the full development and life cycle management of process for all health care products
- Offer a holistic and integrated approach to the development life cycle management of regulated health care products

DIA 2014 50th Annual Meeting provides the best opportunity to meet with people from around the world, share your views and knowledge, network, and build new relationships.

Who Should Attend?

This meeting is designed for individuals involved in the discovery, development, and life cycle management of pharmaceuticals, biotechnology, medical devices, and related medical products. The Annual Meeting is intended to strengthen professionals' understanding of the value of cross-discipline integration and to foster innovation for better health outcomes.

2. Instructions for grant proposals can be long and detailed, so many start with a concise abstract or executive summary. Figure 10.5 shows the executive summary for Project LAUNCH,[6] the same grant RFA discussed in this chapter. Critique the effectiveness of this executive summary for highlighting important details from an administrator's perspective. (You may want to compare this executive summary with the full RFA.) Would you suggest any revisions? If so, why?

Figure 10.5 Executive Summary for Project LAUNCH Application

EXECUTIVE SUMMARY

The Substance Abuse and Mental Health Services Administration, Center for Mental Health Services is accepting applications for fiscal year (FY) 2014 for Cooperative Agreements for Linking Actions for Unmet Needs in Children's Health (Project LAUNCH). The purpose of Project LAUNCH is to promote the wellness of young children from birth to eight years by addressing the physical, social, emotional, cognitive and behavioral aspects of their development. The goal of Project LAUNCH is to create a shared vision for the wellness of young children that drives the development of federal, state, territorial, tribal, and locally-based networks for the coordination of key child-serving systems and the integration of behavioral and physical health services. The expected result is for children to be thriving in safe, supportive environments and entering school ready to learn and able to succeed.

Funding Opportunity Title:	Cooperative Agreements for Linking Activities for Unmet Needs in Children's Health
Funding Opportunity Number:	SM-14-004
Due Date for Applications:	March 3, 2014
Anticipated Total Available Funding:	$10,202,000
Estimated Number of Awards:	Up to 13
Estimated Award Amount:	Up to $800,000 per year
Cost Sharing/Match Required	No
Length of Project Period:	Up to 5 years
Eligible Applicants:	Eligible applicants are state and territorial governments; federally recognized American Indian and Alaska Native (AI/AN) Tribes and Tribal organizations. Current and previously funded state and tribal Project LAUNCH grantees are not eligible to apply. [See Section III-1 of this RFA for complete eligibility information.]

3. While many of us do not routinely write documents *for* administrators, we often need to respond to requests *from* administrators, whether insurance firm, medical office, or government agency. Some simply request information; others may deny a claim for reimbursement. For one such request that you've received, prepare a response. This time, though, consider the position of the healthcare administrator as you prewrite. How should you construct this document to fulfill the needs and expectations of your audience?

References

1 Substance Abuse and Mental Health Services Administration. *Applying for a New SAMHSA Grant*. Rockville, MD: Substance Abuse and Mental Health Services Administration; last updated December 19, 2013. http://beta.samhsa.gov/grants/applying.

2 Substance Abuse and Mental Health Services Administration. *Cooperative Agreements for Linking Actions for Unmet Needs in Children's Health.* [RFA Number SM-14-004.] Rockville, MD: Substance Abuse and Mental Health Services Administration; posted December 26, 2013; page 25. http://beta.samhsa.gov/grants/grant-announcements/sm-14-004.

3 Medicare Learning Network. *Therapy Caps Exception Process.* Baltimore: US Centers for Medicare and Medicaid Services; last updated November 1, 2012. MLN Matters Number MM4364. http://www.cms.hhs.gov/MLNMattersArticles/downloads/MM4364.pdf.

4 Task Force on Resource Mobilization and Management Strategies. *Task Force on Resource Mobilization and Management Strategies—Final Report*. Geneva: World Health Organization; 2013. http://apps.who.int/about/who_reform/TFRMMS-report-2013.pdf.

5 Drug Information Association. *DIA 2014 50th Annual Meeting: Celebrate the Past—Invent the Future.* Washington, DC: DIA Global; February 2, 2014. http://www.diahome.org/en-US/Flagship-Meetings/DIA2014/.

6 Substance Abuse and Mental Health Services Administration. *Cooperative Agreements for Linking Actions for Unmet Needs in Children's Health.* [RFA Number SM-14-004.] Rockville, MD: Substance Abuse and Mental Health Services Administration; posted December 26, 2013; page 4. http://beta.samhsa.gov/grants/grant-announcements/sm-14-004.

11

SCIENTIFIC DUALITY OF PRACTITIONERS

Chapter Objectives

In this chapter, you will learn to:

- Comprehend the dual nature of the practitioner audience
- Target the practitioner's acuity for scientific knowledge
- Facilitate hands-on use through strategies and mechanics
- Design documents with formatting that promotes access

The Duality of the Practitioner Audience

In the previous two chapters, we've seen that laypersons come in a wide variety that necessitates envisioning a composite, and that administrators with their diverse roles all share a unique focus. Since we've decided *not* to discuss researchers, we've now come to our last audience group—practitioners.

The practitioner audience shares characteristics with administrators and researchers. Like administrators, practitioners emphasize practicability, which simply means the ability to put into practice. In other words, this group seeks to accomplish real tasks in real time for real outcomes. The difference between practitioners and administrators derives from what practitioners share with researchers: an acuity for medical, healthcare, and scientific knowledge—rather than business and management. While both groups (i.e., practitioners and researchers) often have the same technical knowledge, the distinction for writers is use of that knowledge: application vs. exploration.

Respecting Scientific Knowledge

Let's begin with the scientific side of our practitioner audience. Members of this group fill an increasingly diverse set of roles, reflecting the increasingly diverse types of specialization in today's healthcare arena. As noted in Chapter 2, the traditional roles of practitioners include family physicians, who treat patients; registered nurses, who tend the ill; medical technologists, who analyze diagnostic samples; and retail pharmacists, who counsel about drugs. To meet changing needs, roles have expanded to encompass physician assistants, who work in tandem with doctors; nurse practitioners, who handle routine healthcare issues; clinical pharmacists, who oversee drug delivery in hospitals; and instrumentation technicians, who specialize by area.

Clearly, documents for practitioners must target a specific subgroup, since each has its own accepted jargon. Terminologies for those who perform magnetic resonance imaging (MRI), for instance, would include abbreviating symbols, units of measurement, and numerical settings. Writers would not need to define basic terms or explain routine procedures for MRI technologists. Also, writers might exclude interpretive details for the technologist, since actual diagnosis would fall within the role of a supervisory physician (i.e., a practitioner of another subgroup).

Tailoring for Application

Because practitioners want to *apply* knowledge, documents for practitioners must be easily usable. Envision a medical technologist in a hospital lab: he doesn't have

Table 11.1 Scenarios of Purpose, Content, and Strategies for Practitioners

PRACTITIONER ROLES	DOCUMENT PURPOSE
Pediatric Physician or Nurse Practitioner	To remind about age for administration of vaccinations, both required and recommended, for infants
Intern or Resident in Emergency Setting	To delineate procedures for triage that allow handling patients by level of severity of their conditions
Clinical Pharmacist assisted by Nursing Supervisor	To set dosing regimens for patients' drugs, including flagging of possible drug-drug interactions
Medical Technologist supervised by Clinical Pathologist	To explain new instrumentation, as well as to train personnel on proper operation and maintenance

Note: These document examples can apply to more than one practitioner group, given overlap of responsibilities.

time to peruse pages of text to troubleshoot a recurring problem for a piece of instrumentation. Or consider a clinical pharmacist whose patients already take various prescription drugs, along with those being added for treatment in the hospital: she can't spend time thumbing through reference tomes. Those documents serve another purpose for another audience—not these practitioners.

Practitioners, as in these two hypothetical examples, crave information presented for quick access (as already seen for administrators in Chapter 10). Techniques such as bulleted or numbered steps, organizational headings for hierarchies, and appendices for ancillary materials work well. Strategies and mechanics come next.

Techniques for Practitioner Audiences

Strategies that Balance the Duality of Roles

Before discussing specific strategies, let's first revisit the issues of *purpose* and *content* for documents targeting practitioners. Because practitioners fulfill a diversity of functions, writers for this audience group must understand precisely what each document intends to achieve (purpose) and what materials it must include (content). Table 11.1 lists hypothetical scenarios for the purpose and content of practitioner-focused documents. Given the overlap of responsibilities among practitioners, some documents may apply to more than one practitioner group or subgroup.

In the first scenario, the timing of vaccinations is critical, so a chronology strategy makes sense; a timeline (table or graphic) should complement the text. Writers also have two groups to consider, so further classification could be by disease to be prevented and by whether the vaccination is required vs. recommended.

DOCUMENT CONTENT	DOCUMENT STRATEGIES
Specific vaccinations, with dosages and timing, as well as potential adverse events	Chronology of vaccinations; classification by disease, as well as required vs. optional
Decision points defining the level of severity, along with procedures for admittance	Order of importance nested by severity and condition; flow chart for procedures
Patient-specific drug listing with dosages and timings, as well as warning signs	Chronological drug ordering; partition of warning signs by drug class and severity
Overview of instrumentation, with operating principles, as well as working procedures	Specificity for overview with cause-and-effect and spatiality; chronological order of steps

Practitioners do not require the procedure for administering vaccinations—that might insult this audience.

The second scenario, emergency triage, does need admitting procedures, perhaps succinctly presented as a flow diagram. Since these practitioners must determine which patients receive priority, an importance strategy would be appropriate, most likely as a nested strategy within an easy-to-access guide.

For the third scenario of hospital drugs, the clinical pharmacist would be the main audience, although the nursing supervisor ensuring implementation of the regimen would be an important secondary audience. In this case, procedures would include a chronology regarding the administration of specific drugs. Given the potential for drug interactions, warning signs might be partitioned by drug class and then classified by severity.

Finally, the fourth scenario on instrumentation targets a laboratory technologist but also recognizes the clinical pathologist heading the lab as a secondary reader. Since the technologist might not be familiar with this new instrument, an overview of operating principles should precede more specific details. For both operation and maintenance, the technologist would want the cause-and-effect relationships of operation and troubleshooting. Spatially organized visuals could show the instrument's internal workings, complemented by numbered steps for the procedures themselves.

Mechanics that Facilitate Practical Use

With audience, purpose, and content addressed, writers for these hypothetical scenarios (Table 11.1) next need to consider writing mechanics. Several of the scenarios involve chronological steps, which could be numbered. These steps typically would be written in imperative (i.e., command) mood for clarity. However, some steps might need additional material: warning flags, clarifications, options, and so on. Such additional text might not flow as well in imperative mood, so declarative mood might suffice. Moreover, simpler syntax works well for procedural steps, but subtleties might require complex or compound structures. Writers could highlight such text in boxes and with special fonts or colors. For these highlighted statements, active voice would achieve greater clarity than would passive constructions.

Given the need for clarity to facilitate use, care must be taken with restrictive vs. non-restrictive clauses, phrases used for description, and pronouns to avoid long strings of nouns. The type of relative clause should reflect the importance of the conveyed information. Descriptive phrases, especially those beginning with participles, should be placed next to terms being modified to avoid confusion. Because vagueness can obfuscate meaning, the antecedents for any pronouns, especially *it*, must be obvious.

Diction, or word choice, for practitioner audiences must be precise. The correct term should be used each time; synonyms simply for variety might introduce ambiguity, thereby detracting from ease of use. Definitions should not be needed, but an appended glossary of new terms would assist less-versed individuals. Above all, vocabulary does not need dilution. Whether monosyllabic or polysyllabic, the correct term should be used (and abbreviated, if used repeatedly).

Document Examples for Practitioners

The same considerations of purpose, content, strategy, and mechanics apply both to hypothetical and real scenarios. The following three authentic examples include an overview of angina pectoris diagnosis, protocol procedures for a clinical trial, and training materials used in public health.

Diagnostic Overview of Angina Pectoris

First-line diagnosticians—including physicians, physician assistants, and nurse practitioners—maintain a wealth of medical information in their brains. Regardless, practitioners occasionally must consult references to aid in the diagnosis of conditions that may be rare, complicated by sequelae (i.e., concomitant conditions), or symptomatically like another disease. From one of these well-respected guides, *The Merck Manual of Diagnosis and Therapy*, comes the first example excerpt, an overview of angina pectoris to assist practitioners during diagnosis (Figure 11.1).[1]

The format of this excerpt is straight text. Headings partition the text into usable sections for quick scanning by busy practitioners. The italicized font draws attention to an extended definition (for those who need clarity of this condition). Because knowledge of cardiac physiology is assumed, medical terms are not defined. These polysyllabic terms are counter-balanced by relatively straightforward syntax.

Of the 12 sentences within the excerpt, eight (two-thirds) are simple. Given the focus on causes and effects of angina pectoris, three sentences are complex. This passage even has one compound-complex sentence (last sentence of first paragraph of "Etiology"); however, this short sentence (17 words) poses no difficulties. The complex sentence comprising the second paragraph of "Etiology" uses a cause-effect strategy for conditions precipitating angina pectoris; its subordinate (dependent) clause relates the narrowing of coronary arteries to angina pectoris induced by exertion.

Figure 11.1 Overview Description (Excerpt) of Angina Pectoris

Angina Pectoris

Angina pectoris is a clinical syndrome of precordial discomfort or pressure due to transient myocardial ischemia without infarction. It is typically precipitated by exertion or psychologic stress and relieved by rest or sublingual nitroglycerin. Diagnosis is by symptoms, ECG, and myocardial imaging. Treatment may include aspirin, nitrates, ß-blockers, Ca channel blockers, ACE inhibitors, statins, and coronary angioplasty or coronary artery bypass graft surgery.

Etiology

Angina pectoris occurs when cardiac workload and resultant myocardial O_2 demand exceed the ability of coronary arteries to supply an adequate amount of oxygenated blood, as can occur when the arteries are narrowed. Narrowing usually results from atherosclerosis but may result from coronary artery spasm or, rarely, coronary artery embolism. Acute coronary thrombosis can cause angina if obstruction is partial or transient, but it usually causes MI.

Because myocardial O_2 demand is determined mainly by heart rate, systolic wall tension, and contractility, narrowing of a coronary artery typically results in angina that occurs during exertion and is relieved by rest.

In addition to exertion, cardiac workload can be increased by disorders such as hypertension, aortic stenosis, aortic regurgitation, or hypertrophic cardiomyopathy. In such cases, angina can result whether atherosclerosis is present or not. These disorders can also decrease relative myocardial perfusion because myocardial mass is increased (causing decreased diastolic flow).

A decreased O_2 supply, as in severe anemia or hypoxia, can precipitate or aggravate angina.

Excerpt from The Merck Manual of Diagnosis and Therapy, Porter RS, ed. © 2010-2013 Merck Sharp & Dohme Corp., a subsidiary of Merck & Co, Inc., Whitehouse Station, NJ. Reprinted with permission. http://www.merckmanuals.com/professional/. Accessed May 22, 2014.

In the excerpt, no sentences are compound; however, these structures appear later in the document, such as in "Symptoms and Signs": "Frequently, the patient's description is imprecise, and whether the problem is angina, dyspnea, or both may be difficult to determine." Interestingly, this section identifies those patients *without* angina pectoris: "Because discomfort seldom occurs in the region of the cardiac apex, the patient who points to this precise area or describes fleeting, sharp, or hot sensations usually does not have angina." Negation to restrict a group is often an underrated writing technique that many writers neglect.

Also, that sentence uses a complex structure; its subordinate clause states where pain seldom occurs so that the main clause can emphasize the exclusion of patients who indicate pain in that specific area. This same independent clause includes an important restrictive clause: only the individual "who points to this precise area" would usually be excluded from a diagnosis of angina pectoris. Overall, the document typifies how choices of syntax reinforce content for target audiences.

Protocol Procedures for Clinical Trial

Figure 11.2 takes this text approach for practitioners a step further through format and design in a clinical protocol examining the respiratory infection SARS (Severe Acute Respiratory Syndrome).[2] Essentially, clinical protocols detail procedures for conducting trials of vaccines, drugs, and other treatments in humans. Trials can range from small scale (i.e., 10 healthy volunteers, one drug, a few days) to large scale (i.e., thousands of patients with the disease being studied, different dosages of the test drug and comparative agents, and lasting years). Because the directing research physician relies on monitors for the daily handling of compared treatments, evaluative tests, and massive paperwork, the protocol (trial plan) must meet the operational needs of this secondary audience. Moreover, clinical research trials must adhere to stringent requirements so that data can be analyzed while ensuring that participants' well-being is respected.

Figure 11.2 Clinical Protocol for a Study of SARS

SARS APHERESIS SCREENING—INCLUSION CRITERIA (ALL OF THE FOLLOWING):

1. SARS Ab level greater than or equal to 1:80
2. Adequate clinical parameters (must meet all of the criteria below):
    1. Afebrile (temperature less than 38 degrees C)
    2. Systolic blood pressure greater than 100 and less than 160 mmHg
    3. Diastolic blood pressure greater than 50 and less than 100 mmHg
    4. Heart rate between 50-100 beats/minute
    5. Weight greater than or equal to 110 pounds (50kg)
3. Adequate laboratory parameters (must meet all of the criteria below):
    1. Hemoglobin greater than or equal to 12.5 g/dL for females or 13 grams/dL for males
    2. HCT greater than or equal to 38%
    3. Platelets greater than or equal to 150,000 x 10(3)/microL
    4. Total serum protein greater than or equal to 6.0 grams/100 milliliters of blood
    5. ALT less than 1.5 Upper limit of normal
    6. Quantitative immune globulin levels (all of the following):
        - Ig A greater than or equal to 90 mg/dL
        - Ig G greater than or equal to 650 mg/dL
        - Ig M greater than or equal to 35 mg/dL
4. Negative tests for blood borne pathogens (must meet all of the criteria below):
    1. Human Immunodeficiency Virus 1/2 Antibody (Anti-HIV1/2) negative
    2. Hepatitis B virus surface antigen (HBsAg) negative
    3. Hepatitis C virus antibody (anti-HCV) negative
    4. RPR non-reactive
    5. HTLV antibody
5. Adequate peripheral venous access for plasmapheresis (as judged by the examiner)

This second example lists inclusion criteria for a patient to be admitted into the research trial. In addition to text descriptions for the primary audience of the directing physician, a tabular format encapsulates these details for the secondary audience of the trial's assisting monitor. The protocol's format uses generous indentations to highlight numbered hierarchical criteria for enrollment eligibility. These criteria are expressed as phrases for quick reference. One can imagine how such lists could link to forms for recording the requisite clinical and laboratory data; furthermore, this checklist format would familiarize other personnel working on this same trial in a supportive role. However, anyone directly using this information would need familiarity with its many technical terms.

Training Materials for Public Health

The third example (Figure 11.3) combines text with a table as an online tutorial for practitioners in the field of public health.[3] Unlike practitioners who treat individuals, practitioners in public health must look at the health of society overall. For example, a physician in a college's health center may treat a student with meningitis (a recurring problem), but a physician in the regional office of public health must examine the risk of this communicable disease being spread widely to other students and, if the student is not isolated, potentially beyond the campus.

As shown in Figure 11.3, these busy practitioners must stay aware of all relevant healthcare information for their community or group. An organization called Partners in Information Access for the Public Health Workforce provides an online tutorial to assist these practitioners in staying informed. This excerpt from the tutorial has two purposes: a primary purpose to train the practitioners, with a secondary purpose to persuade them of the importance to add this task to their busy schedules.

This excerpt starts with an overview on the importance of staying informed before giving specific persuasive arguments for doing so. In the overview, two of the four sentences are compound, yet the paragraph does not sound difficult when read. Because this paragraph compares public health with business, the compound structure of the second sentence sets both entities as equals; this technique reinforces the persuasive purpose. The fourth sentence again uses a compound structure to emphasize the nature of communication as "a two-way street." The ensuing reasons for staying informed partition into answers to the question "Why adopt a new strategy for staying informed?" These statements use second person to directly address practitioners. Not only is it easy to follow, but this grammatical structure also reinforces the document's persuasive purpose.

As can be seen with these three examples for the audience groups, healthcare is a critical field in which information must be conveyed efficiently and effectively. The final chapters of this book now turn first to an exploration of how to fine-tune and finalize healthcare documents through revision, and then to the ethics and legalities at the heart of healthcare writing.

Figure 11.3 Online Tutorial about Information and Data for Public Health

Public Health Information & Data Tutorial: Staying Informed

While collaborative in nature, public health agencies and community based organizations often find themselves competing with industry for the public's attention and funding for health concerns. Businesses describe staying informed as competitive intelligence, and like businesses, public health organizations cannot afford to be out-of-date. Staying informed about news and developments in public health is a crucial component of health competencies related to communication. Communication is a two-way street, and public health practitioners need to use the media, advanced technologies, and community networks to gain information as well as communicate it.

Why adopt a new strategy for staying informed?

Professional responsibility	The Public Health Code of Ethics suggests that public health workers should seek the information needed to implement effective policies and programs that protect and promote health.
Identify experts in the field	Watching the news in public health and knowing who publishes on topics of interest can help you identify experts in other agencies, academia, and industry.
Career development & advancement	You want to be aware of the latest information, so you can answer, "Yes!" when a colleague asks you if you heard about an important paper in a recent journal. Being informed about developments in public health is important at every level of your career.
Economic forces	It may not be economically feasible for you to go to regional or national meetings to keep informed. In difficult budget times, you may be unable to travel to as many meetings or attend as many continuing education offerings.
Have information as soon as it's available	A working strategy makes it possible to have information as soon as it's available—no waiting for a subscription to print materials or interoffice routing.

Chapter Summary

- Practitioners, like researchers, have a strong command of the scientific knowledge in their fields. Unlike researchers, though, practitioners focus more on application of medicine and healthcare, rather than exploratory research in these areas.
- Documents for practitioners need to respect their scientific acuity. Definitions, for instance, are usually not needed but can be appended just in case.
- Writing strategies should consider that practitioners directly use information, so unwieldy text is not advisable. Instead, documents should be clearly organized, with subsections using nested strategies. Sentences are often structured concisely, although syntax other than simple may still be applicable.
- To facilitate comprehension and application, documents should be formatted with the practitioner's focus in mind. Bulleted lists, numbered steps, generous margins, and complementary graphics are format techniques that complement the text.

Exercises for Practice

1. Medical instrumentation that assists healthcare practitioners in diagnosing their patients' conditions seems to advance every day. Given such change, practitioners need to remain up-to-date on which techniques apply to suspected conditions, as well as which techniques may have become obsolete. Figure 11.4 overviews two techniques for visualizing conditions suspected of stemming from pulmonary disorders.[4] Read this excerpt with a focus on the dual nature of practitioners. Describe the audience group in terms of technical knowledge and information use. Which characteristics of this document support your description?

Figure 11.4 Text-Based Overview (Excerpt) of Two Methods for Chest Imaging

Magnetic Resonance Imaging

MRI has a relatively limited role in pulmonary imaging but is preferred over CT in specific circumstances, such as assessment of superior sulcus tumors, possible cysts, and other lesions that abut the chest wall. In patients with suspected pulmonary embolism in whom IV contrast cannot be used, MRI can sometimes identify large proximal emboli but usually is limited in this disorder. The use of MRI to evaluate pulmonary hypertension is being studied, and this practice may become more common.

Advantages include absence of radiation exposure, excellent visualization of vascular structures, lack of artifact due to bone, and excellent soft-tissue contrast. Disadvantages include respiratory and cardiac motion, the time it takes to do the procedure, and the occasional presence of absolute or relative contraindications.

Ultrasonography

Ultrasonography is often used to facilitate procedures such as thoracentesis and central venous catheter insertion. Endobronchial ultrasonography (EBUS) is increasingly being used in conjunction with fiberoptic bronchoscopy to help localize masses and enlarged lymph nodes. Diagnostic yield of transbronchial lymph node aspiration is higher using EBUS than conventional unguided techniques. Ultrasonography is also very useful for evaluating presence and size of pleural effusions and is now commonly used at the bedside to guide thoracentesis.

Excerpt from The Merck Manual of Diagnosis and Therapy, Porter RS, ed. © 2010-2013 Merck Sharp & Dohme Corp., a subsidiary of Merck & Co, Inc., Whitehouse Station, NJ. Reprinted with permission. http://www.merckmanuals.com/professional/. Accessed May 22, 2014.

2. Viruses range from simple ones that cause the common cold to malicious ones that result in excruciatingly painful death. An example of the latter is the Ebola virus. Infection by this virus leads to hemorrhagic fever, which can produce not only mild symptoms, such as muscle aches, but also severe internal and external hemorrhaging. Because the subset of patients who die includes those who have not mustered sufficient immunologic response, vaccination may be one means to halt the spread of this contagious disease. The text in Figure 11.5 details the purpose of a clinical trial studying a vaccine to be used against the Ebola virus.[5] Using this information, construct a table or figure (such as a flow chart) to complement the text; remember that the audience could be a physician, nurse, or other healthcare worker already knowledgeable about viral infections.

Figure 11.5 Protocol for Clinical Study of Ebola Virus Vaccine

This study will test the safety of an experimental vaccine developed to protect against Ebola virus infection and to determine if the vaccine induces an immune response to the virus. The Ebola virus causes a disease called Ebola hemorrhagic fever. Symptoms begin with fever and muscle aches and progress to breathing problems, severe bleeding, kidney problems, and shock. The infection may be mild, but it can also lead to death. The vaccine used in this study is made from small parts of Ebola genetic material. It cannot cause Ebola hemorrhagic fever to develop in those who receive it.

Healthy volunteers 18 to 44 years of age may be eligible for this study. Candidates will be screened with a medical history, physical examination, and laboratory tests, and complete an "assessment of understanding" questionnaire to show that they understand the study.

Depending on their order of entry into the study, participants are assigned to receive one of three vaccine doses or placebo. The first group receives the lowest dose (2 milligram) of vaccine or placebo. If this dose is safe, then the second group receives 4 mg, and if this dose is safe, the third group receives 8 mg. Injections are given in a muscle in the upper arm. Participants receive three injections, each 4 weeks apart (on study days 0, 28, and 56).

Participants record their temperature and symptoms in a diary card for 7 days following each injection. They return to the clinic 2 to 3 days after each injection and then 2 weeks after each injection until study week 10. Additional follow-up visits are then scheduled at weeks 12, 24, 38, and 52. At each visit, participants provide a blood and urine sample for testing and have their vital signs, and lymph nodes checked, their weight measured, and their symptoms reviewed. Additional laboratory tests may be requested between visits.

3. Outbreaks of hemorrhagic fever caused by the Ebola virus have recently occurred in West African countries, notably Guinea, Liberia, Mali, Nigeria, Senegal, and Sierra Leone. More recently, however, some cases have occurred in the US and potentially other Western nations through the travel of healthcare workers. The Centers for Disease Control therefore created an online document on this infection, with the main text structured in a question-and-answer format; Figure 11.6 provides an excerpt.[6] Identify strategies, mechanics, and other writing techniques used in this information packet. Does it target the same type of healthcare practitioner as in the previous exercise? How do you know?

Figure 11.6 Question-and-Answer Format for Ebola Hemorrhagic Fever

What are the symptoms of Ebola hemorrhagic fever?

The incubation period for Ebola HF ranges from 2 to 21 days. The onset of illness is abrupt and is characterized by fever, headache, joint and muscle aches, sore throat, and weakness, followed by diarrhea, vomiting, and stomach pain. A rash, red eyes, hiccups and internal and external bleeding may be seen in some patients. Researchers do not understand why some people are able to recover from Ebola HF and others are not. However, it is known that patients who die usually have not developed a significant immune response to the virus at the time of death.

How is Ebola hemorrhagic fever clinically diagnosed?

Diagnosing Ebola HF in an individual who has been infected only a few days is difficult because early symptoms, such as red eyes and a skin rash, are nonspecific to the virus and are seen in other patients with diseases that occur much more frequently. However, if a person has the constellation of symptoms described above, and infection with Ebola virus is suspected, isolate the patient and notify local and state health departments and the CDC.

What laboratory tests are used to diagnose Ebola hemorrhagic fever?

Antigen-capture enzyme-linked immunosorbent assay (ELISA) testing, IgM ELISA, polymerase chain reaction (PCR), and virus isolation can be used to diagnose a case of Ebola HF within a few days of the onset of symptoms. Persons tested later in the course of the disease or after recovery can be tested for IgM and IgG antibodies; the disease can also be diagnosed retrospectively in deceased patients by using immunohistochemistry testing, virus isolation, or PCR.

How is Ebola hemorrhagic fever treated?

There is no standard treatment for Ebola HF. Currently, patients receive supportive therapy. This consists of balancing the patient's fluids and electrolytes, maintaining their oxygen status and blood pressure, and treating them for any complicating infections.

Note: Minor typographical and formatting changes were made to this document.

4. The US Food and Drug Administration (FDA) strictly monitors the safety of all prescription drugs on the US market. Pharmaceutical manufacturers must follow requirements for reporting known instances of adverse events that occur with their products. Practitioners, such as physicians, are not mandated to report any adverse events (i.e., side effects) that their patients experience, but the FDA encourages a voluntary reporting system. Classification of any adverse events, including their category and severity, is critical to this process; to that end, the FDA employs a specific classification system. Figure 11.7 provides an excerpt from instructions for completing the online reporting form.[7] Which writing techniques did the FDA employ? How broad is the audience who might report adverse events of patients? Does this breadth affect the document? If so, how? Which changes would you suggest if the FDA were to revise this form? Why?

Figure 11.7 Classification of Adverse Events for Prescription Drugs

Adverse event: Any incident where the use of a medication (drug or biologic), at any dose, human cell/tissue product, blood product, a medical device (including *in vitro* diagnostics) or a special nutritional product (e.g., dietary supplement, infant formula or medical food) is suspected to have resulted in an adverse outcome in a patient.

To report, it is not necessary to be certain of a cause/effect relationship between the adverse event and the use of the medical product(s) in question. Suspicion of an association is sufficient reason to report. Submission of a report does not constitute an admission that medical personnel or the product caused or contributed to the event. For whole blood and blood components, adverse events can include adverse outcomes in a blood donor or recipient.

Please limit your submissions to those events that are serious. An event is classified as serious when the patient outcome attributed to the adverse event is:

- Death
- Life-threatening
- Hospitalization (initial or prolonged)
- Disability or permanent damage
- Congenital anomaly/birth defect
- Required medical or surgical intervention to prevent permanent impairment or damage (devices)
- Other serious (important medical events)

Note: Minor typographical and formatting changes were made to this document.

References

1 Angina pectoris. In: Porter RS, ed. *The Merck Manual of Diagnosis and Therapy.* Whitehouse Station, NJ: Merck Sharp & Dohme; 2010-2013; last full review/revision May 2013; last modified October 2013. http://www.merckmanuals.com/professional/.

2 US National Institutes of Health Clinical Center. *Collection of Convalescent SARS Plasma by Apheresis.* Bethesda, MD: US National Institutes of Health; last verified February 2007. [Clinical protocol]. ClinicalTrials.gov Identifier NCT00342524. http://www.clinicaltrials.gov/.

3 Partners in Information Access for the Public Health Workforce. *Public Health Information & Data Tutorial.* Bethesda, MD: National Library of Medicine; last updated January 4, 2013. http://www.phpartners.org/tutorial/.

4 Chest imaging. In: Porter RS, ed. *The Merck Manual of Diagnosis and Therapy.* Whitehouse Station, NJ: Merck Sharp & Dohme; 2010-2013; last full review/revision June 2013; last modified September 2013. http://www.merckmanuals.com/professional/.

5 US National Institutes of Health Clinical Center. *Experimental Ebola Vaccine Trial.* Bethesda, MD: US National Institutes of Health; last verified August 2007. [Clinical protocol]. ClinicalTrials.gov Identifier NCT00072605. http://www.clinicaltrials.gov/show/NCT00072605.

6 Centers for Disease Control and Infection. *Ebola Hemorrhagic Fever: Information Packet.* Atlanta: US Department of Health and Human Services; first published October 8, 2002; last verified August 2007. http://www.cdc.gov/ncidod/dvrd/spb/mnpages/ebola.pdf.

7 MedWatch. *Voluntary Reporting.* Silver Spring, MD: US Food and Drug Administration; last updated February 24, 2014. Form FDA 3500. http://www.fda.gov/medwatch/getforms.htm.

12

REVISION TO FINALIZE DOCUMENTS

Chapter Objectives

In this chapter, you will learn to:

- Perform editing at both the micro and macro levels
- Apply cohesion and coherence in long documents
- Use discretion with tools that diagnose our writing

Challenges of the Postwriting Stage

Finalizing documents in the postwriting stage relies on the cascade so far covered in this text. After a thorough job during the prewriting stage, the writing stage proceeds more smoothly. Time and effort spent up-front to determine audience, purpose, and context allow an identification of the document's medium, content, and strategy. With strategies as outlines, information sources can generate the right content. As the writing stage proceeds, content can be developed through techniques for argumentation, tables and graphs, and writing mechanics. Given this worthwhile effort, a document should move more quickly during postwriting, the stage of revision and finalization.

This postwriting stage involves the following three editorial challenges:

- Micro vs. macro levels of editing
- Integrity within long documents
- Software tools that diagnose text

Microediting vs. Macroediting

While writers recognize the need for editing documents during postwriting, not all writers may agree on which elements constitute editing. In fact, some mistakenly contend that editing is little more than proofreading for overt mistakes in spelling or punctuation. True, those errors need to be corrected. However, writing involves more than spelling and punctuation—and editing does, too. A critical distinction concerns the level of editing. From what perspective are we peering down on our (or a colleague's) document? Borrowing a science metaphor, we can view not only from our usual vantage, but also through a lens that magnifies the smaller portions of the document:

- **Microediting** focuses on smaller elements of a document, from spelling and punctuation through grammar and syntax
- **Macroediting** focuses on a document as a whole, from organization and headers through format, layout, and design

Documents in technical fields like healthcare contain so many elements (including content) that editing also includes checking data consistency, referencing source materials, calculating numerical extrapolations, and consistently using abbreviations or acronyms.[1] For example, do data in the tables and figures correspond to explanations in the text? Are percentages calculated correctly from original data? Do measurement units match across the entire document, including graphics? Are sources cited correctly?

To check these various editing elements, remember to return to guidelines already discussed: references (Chapter 5), argumentation (Chapter 6), visuals (Chapter 7), and mechanics (Chapter 8). Surprisingly, microediting—albeit tedious—may be the easier level of editing to perform. Some writers (and editors) keep lists of editorial points handy, along with longer style guides. The more difficult aspect of microediting derives from our familiarity with our own documents. To avoid missing errors, try to let a few days elapse between writing and editing. Better yet, ask a colleague to edit your document (and return the favor). We'll see later in this chapter that automated tools, if used with discretion, can aid microediting.

What about macroediting? Governing the structural integrity of documents at this overall level are deliberate decisions made during the prewriting stage as related to input conditions (audience, purpose, and context) and output parameters (medium, content, and strategy). Does the nearly finished document fulfill the expectations developed during the prewriting stage? If not, make adjustments as needed. Signposting of organization, along with headers and subheaders, should signal the document's structure, especially for long documents with nested strategies.

Cohesion and Coherence

Longer documents that require multiple and nested strategies challenge writers in one additional way: how to keep the document fully understandable to the audience? As already discussed, signposting the organization is one important way. Writers, though, must also ensure that the entire document reads as a comprehensible whole—rather than as a piecemeal concatenation that appears disjointed across sections. Composing longer and interrelated sections of documents requires that writers fortify the document through *cohesion* and *coherence*:[2]

- **Cohesion**—Resistance to separation because of internally consistent and related elements that adhere subsections together
- **Coherence**—Structure whose elements are presented logically, orderly, and consistently to read as a single unified whole

Cohesion and coherence reflect how well writing (particularly in longer and more complex passages common in today's healthcare topics) hangs together. In other words, readers must easily sense the interconnectedness and flow of evidence and arguments that are threaded through the document's structure. Certainly, signposting, as already discussed, helps the reader to recognize the organization. Transitional words and phrases (e.g., *however, on the other hand, in addition*) help the reader move from sentence to sentence or paragraph to paragraph. Nonetheless, for information relying on argumentation of greater complexity, an additional layer of linkage fosters cohesion and coherence for the target audience.

Consider these alternative passages:

> *Identifying how viruses are transmitted in our daycare constitutes our main goal. Adherence to processes for washing hands before and after handling children will be measured. A plan for decreasing viral transmission in daycares will be developed from these results.*

vs.

> *Our first goal is to see how viruses are transmitted in our daycare. This goal depends on measuring processes for washing hands before and after handling children. From these results, a plan for decreasing viral transmission in daycares will be developed.*

The same information flows more readily in the second version. Positioning of repeated words ("goal") or similar terms ("measuring processes for washing hands before and after handling children" and "results") bridges one sentence to the next. These techniques can be used to bridge paragraphs to paragraphs, as well as sections to sections, in documents of increasing length and complexity.

Identifying the *true topic* of each sentence or paragraph in a sequence strengthens cohesion and coherence. Surprisingly, the topic may not be the grammatical subject of a sentence, especially those more varied in syntax. For example, consider this sentence:

> *Regarding our results, their applicability to other daycares may be limited because of the age distribution within our particular center.*

Which of these several underlined terms is the true topic? Remember that the true topic internally links the full paragraph, so which of these terms will be the focus of the next sentences? That flow from the true topic to subsequent sentences builds cohesion and coherence for readers.

Similarly, for sections of a long document such as a training manual for employees of the daycare, each paragraph in a section should maintain a reasonable flow from one topic to another. As already discussed, the manual would have a selected strategy; within each section defined by that strategy, individual paragraphs need to focus on their true topic. In some cases, other related topics may fit into the same paragraph, as long as these topics logically flow in sequence within the mind of the reader.

In the daycare example, suppose that the results indicated that the handwashing procedures were not sufficiently emphatic about how often the employees need to wash their hands; furthermore, perhaps the daycare manager feels that the employees would comply more fully if they understood how simple handwashing could greatly decrease the spread of viruses among the children. We might then want two main sections of the document: first, the rationale for the importance of handwashing in decreasing viral transmission; and second, a step-by-step procedure emphasizing frequency and length of washing time. The first part might use an overall strategy of specificity—providing an overview before the specific reasons, each handled with cause-and-effect evidence. The second part could then follow in a chronological order of steps, or perhaps start with the main handwashing factors (e.g., frequency and time) in order of importance. In the latter case, the document would then need a third section on the handwashing process itself.

Experienced writers use these same techniques of enhancing document strategies through cohesion and coherence. Focusing on such techniques may slow down a writer, particularly when facing writer's block. That's why many writers save fine-tuning until the postwriting stage. For many beginning writers, getting the information down within the planned structure is the hardest part of writing. Don't slow yourself down with finer points until you're ready.

Diagnostic Tools for Postwriting

Today's software for document production not only facilitates word processing, but also supplies ancillary tools to assist editing. Although tools vary from one package to another, automated checks of spelling and grammar are both typically provided. Packages with more sophistication add automated assessments of document complexity through readability formulas and syntax analysis. The major problem, as writers realize, lies not with these tools themselves, but with using them appropriately by applying our own judgment and discretion. Anyone should be able to press the buttons to activate the tools; the astute writer knows how to capitalize on the results.

Automated Checks of Spelling and Grammar

Let's admit it: we all use spell-checker (as we should). All too often, as we type, our nimble fingers commit a mechanical faux pas. However, what we should *not* do is set spell-checkers to correct spelling automatically. Spell-checkers will find and highlight annoying slips like *livre* for *liver* (unless, of course, we're writing in French!). That squiggly red underlining alerts us to such problems. What spell-checkers won't do is highlight an error that actually forms a real word. For instance, *livery* would not be detected, but this mistake could damage a writer's credibility.

More serious is the annoyance that occurs with using standard spell-checkers for specialized documents. Terms common in the medical lexicon might not appear in standard spell-checkers. Although *jaundice* would, *hyperbilirubinemia* would not be recognized by a standard software spell-checker. Consider the purchase of available software that recognizes medical terminology.

Grammar-checkers can be even more problematic. Not all issues highlighted by these automated tools require the suggested changes to be made. Sentences with more complicated syntax, for example, might confuse the software. A subject and its predicate may be separated by intervening phrases whose nouns might trigger a disagreement in number, as in this example:

> *The patient* [singular] *whose clinical-laboratory results* [plural] *indicated kidney problems* [plural] *was* [singular, not the plural "were"] *hospitalized overnight for additional examination.*

Regardless of such anomalies, grammar-checkers can detect the more obvious errors that we are all prone to commit. Just use them judiciously.

Some grammar-checkers also detect finer points, such as the percentage of sentences using passive vs. active voice. As discussed in Chapter 8, active voice generally produces clearer writing by focusing on the true subject of the sentence, without requiring helping verbs and prepositional phrases. At times, though, writers may deliberately want to use passive voice. Laypersons may be more interested in knowing that "effective treatments have been developed" (passive) than "Company X developed effective treatments" (active) for a serious condition. Rather than spotlighting the treatments, active voice in this latter sentence spotlights the company name (which may, of course, be appropriate for an administrator). Again, think before allowing software to make all such changes.

Depending on their personal working style, writers also must decide whether to address each potential error as typed, or to wait until a draft version of the information has been fully committed to paper. Choose the process that works best for you; either one can produce the same overall results—as long as writers supplement these tools with their own professional knowledge.

Analysis of Syntax and Readability

Other software tools provide an editorial diagnosis of the document in terms of syntax and readability. As with spell-checkers and grammar-checkers, writers must use discretion with these tools.

Consider the "word count" tool, which can be set to run automatically along with spell-checkers and grammar-checkers. In addition to the number of words in a document (or selected portion), counts of sentences and paragraphs usually appear. The word count can be a useful guide for monitoring document length, particularly for documents limited by size. Counts of sentences and paragraphs also provide a sense of the overall structure of the document. For instance, a document of many words but few sentences may require syntactic adjustment, especially for laypersons.

Advanced software tools take these basic counts one step further by calculating the readability level of a document. While the various formulas for readability are beyond the scope of this chapter, these tools can be used to determine the approximate education needed to understand a document. A level of 12, for instance, indicates that most readers should have a high-school education (or the equivalent) to comprehend the information without excessive difficulty.

Although each readability formula has its own internal algorithm, these formulas typically consider the average number of syllables per word and of words per sentence. (More complex formulas consider how easily a reader can interpolate missing words from the document's overall context.) Such averages are useful in roughly gauging if a document targets a less knowledgeable or a more knowledgeable audience. However, these calculations of readability usually do not take syntax into account. As seen in various examples throughout this book, simple sentences (which are grammatically easier to understand) can confuse readers if the content is too esoteric; conversely, compound-complex sentences (which are more grammatically obtuse) may present concepts clearly and without undue overall length.

These examples illustrate the point:

> *Our technological milieu of healthcare presupposes juxtaposition between provider and patient for efficiency and effectiveness of implementation.*
> [17 words; simple]

vs.

> *Although technology makes great strides, providers and patients must work as partners; this partnership produces better results.*
> [17 words; compound-complex]

Both sentences contain the same number of words. Moreover, the first one uses a simple syntax, whereas the second uses a compound-complex syntax. Nevertheless, the second version is easier to understand, particularly for laypersons.

This example again illustrates that writers must make deliberate decisions for their documents. All the chapters so far have provided the information that writers need to inform their work. In the next (and final) chapter, we'll look at the responsibilities that come with being professionals.

Chapter Summary

- Decisions made during the prewriting stage serve to guide the writing phase so that the postwriting stage proceeds smoothly. During document finalization, time and effort need to be focused on fine-tuning.

- Editing is more than proofreading. The checks that occur during editing include both microediting at the level of writing mechanics and macroediting at the level of structure. Together, both levels of editing strengthen a document.

- Cohesion and coherence are two attributes that reflect a document's quality. Cohesion refers to internal consistency that holds sections together, whereas coherence refers to the overall unity of the document.

- Diagnostic tools can assist writers in fine-tuning their documents provided that suggested changes are reviewed carefully. Checkers for spelling and grammar can often be confused by the intricacies of healthcare information. Readability and other techniques can also indicate the document's potential suitability.

Exercises for Practice

1. Construct your own guideline sheet for editing errors that are the most difficult for you to find. Divide these items between the microediting and macroediting levels. Keep your guideline sheet handy as a checklist for editing your own documents. Continue updating these checklists so that you can chart your personal progress as a healthcare writer and editor.

2. One of the best ways to understand what needs to be covered during postwriting is to review how final documents came to be that way. For this exercise, select one of the example documents from each of Chapters 9, 10, and 11, for the layperson, administrator, and practitioner audiences, respectively. Retype the document text; don't bother with headers, graphics, or other formatting. Then use techniques and tools discussed in this chapter to analyze the three documents in terms of syntax and readability. How well do these documents measure up to the standards that we've discussed in this chapter? What changes might you suggest if you were to revise any of these documents?

3. Investigate the availability of more sophisticated tools dealing with document characteristics such as readability. Research the underlying algorithms used by different formulas. Do any particular formulas seem more appropriate to writing about healthcare? Justify your answer.

References

1 Churchill RE. A gentle introduction to microediting. *AMWA J.* Fall 2000;15(4):19-23. http://www.amwa.org.

2 Williams JM. Cohesion and coherence. In: *Style: Ten Lessons in Clarity and Grace.* 8th ed. New York: Pearson-Longman; 2005. Lesson Five.

13

LEGAL AND ETHICAL RESPONSIBILITIES

Chapter Objectives

In this chapter, you will learn to:

- Appreciate the social contract as entrusted to professionals
- Grasp the application of the legalities of copyright and fair use
- Prepare for evolving responsibilities for healthcare writers

Now that we've reached the final stages of the preparation of healthcare documents, let's step back: document finalization is a worthwhile checkpoint for our professional role as healthcare writers. Issues related to our professional responsibilities to society were first discussed in Chapter 1. Across the prewriting, writing, and postwriting stages, examples and exercises provided opportunities to glimpse these responsibilities in action. Now, in this final chapter, we return to the underlying issues, specifically our legal and ethical responsibilities as healthcare writers.

Legal Responsibilities of Healthcare Writers

Although often intertwined, the legal and ethical responsibilities of any professional can best be understood by reviewing them separately. A professional's social contract, as discussed in Chapter 1, reflects society's expectations for that professional's ethical behavior in serving society; some of these expectations are promulgated as laws or regulations. This section explores three legalities pertinent to the services provided by professional healthcare writers.

Governmental Regulations

Requirements in healthcare can be formally codified as laws, others may be guidelines strongly recommended by regulatory agencies, and some are enforced by professional societies. For instance, the US Food and Drug Administration (FDA) maintains expansive guidelines, not only for research and development of new drugs, but also for the researcher-targeted documents that must be submitted before a new drug can be marketed. While requirements do not apply to documents targeting layperson, administrator, or practitioner audiences, some indeed may: FDA guidelines, for instance, apply to the direct marketing of drugs to laypersons. Healthcare writers must therefore stay abreast of requirements pertinent to any documents that they may be preparing.

Guidelines from the FDA speak to an audience of specialists. Hence, depending on the balance of science in their backgrounds, not all healthcare writers will be able to use them easily. An overview of the main guidelines, though, places them into perspective. The FDA enforces the Federal Food, Drug, and Cosmetic Act, originally passed in 1938 and updated periodically.[1] In fulfilling its role, the FDA issues guidance documents for specific issues, such as advertising. A critical element of direct-to-consumer advertising concerns disclosure of risk so that individuals have sufficient information for making informed decisions in concert with healthcare providers.[2] Recognizing the evolving healthcare environment, another set delimits the promotion of medical products, not by corporate sponsors (i.e., a pharmaceutical company), but by healthcare groups such as pharmacy benefits managers.[3]

Because governments amend legislation and update guidelines, writers share the onus of keeping informed about such changes. Healthcare organizations, whether for-profit or not-for-profit, need to fulfill their legal responsibilities; healthcare writers do, too.

Confidentiality and Informed Consent

We've all bared ourselves—literally and figuratively—when seeking care and counsel in healthcare situations. At the same time, we all expect confidentiality in treatment. That confidentiality extends to writing that might expose individuals through names, photographs, or descriptions. Healthcare writers must respect the trust of confidentiality conferred by society.

Because healthcare publications often refer to individuals, especially when adding pathos through anecdote, the issue of confidentiality is especially important. Anecdotes, for instance, disclose intimate details about an individual's health and body. Healthcare writers must therefore strive to protect that individual's right to privacy and confidentiality even while trying to benefit society through communicating critical information. The AMWA Code of Ethics (introduced in Chapter 1) contains a statement on this very issue:

> Principle 6: Biomedical communicators should respect the confidential nature of materials provided to them. They should not divulge, without permission, any confiden[tial] ... patient information.[4]

The International Committee of Medical Journal Editors addresses this balance for more researcher-oriented publications; those guidelines also inform documents for other audiences.[5] Notably, unnecessary and identifying information—whether words, photos, or other details—must not be made public without the individual's express written consent. Often, this problem can be avoided by referring to the individual by a pseudonym, as in this example: "Suddenly roused from a sound sleep, Alexis (not her real name) gasped for breath. She reported that she never had this sensation before." When not cumbersome, terms like "the patient" or initials rather than a name may also be utilized. The point is that disclosed information of a personal nature can be granted only by that individual. The process of obtaining express written consent in such situations involves explanation to the individual (or guardian for someone unable to make the decision) of the exact nature and extent of that disclosure; a written form providing release must be obtained.

Copyright and Fair Use

Additional legalities critical to the healthcare writer concern the use and substantiation of source materials. Surprisingly for many, citing a reference is not necessarily sufficient in order to preclude plagiarism. Writers must also consider *copyright*—the legal protection afforded to the author of the source material. Copyright does not grant legal exclusivity of the *idea* behind the source, but rather the *expression* of that idea. For example, suppose that a writer published a patient leaflet on safe sex; copyright would not protect the concept of a patient leaflet on safe sex (other healthcare writers could create their own versions). However, copyright would protect the text, layout, and design of that particular version of information on safe sex. In fact, sketches or photographs found on the Internet may themselves be protected. The key to copyright is that the originator has legal protection to control the use of "'original works of authorship' that are fixed in a tangible form of expression."[6]

Although the original US law concerning copyright has changed over time, its basic precepts provide guidance on avoiding lawsuits. Use of source material without permission of the copyright holder depends on four main characteristics: (1) purpose of the use; (2) nature of the copyrighted work; (3) relative portion of work used; and (4) potential effect on market value of the copyrighted work.[7] This allowance is termed "fair use."

Let's consider a hypothetical example of a children's guide for proper use of an inhaler for asthma medications. Perhaps an association of pulmonologists has already published similar guidelines for adults, and we want to base ours on their exemplary material. (1) If we are a not-for-profit support group with no aim to sell our guidelines, then we do not violate the first factor. (2) The source work itself was intended to provide guidance, which agrees with our goal; thus, we do not violate the second factor. (3) The third factor could be an issue, though, depending on how much of the original guidelines we want to mirror. If the pulmonologists' guidelines cover overall management of asthma, then paraphrasing or even quoting a small subsection on inhalers would be allowed; such borrowing from a small document totally focused on inhalers, however, may require consent from that organization. (4) Assuming that the pulmonologists' version was provided to patients at no charge, then our use would not affect its market value. Overall, though, we may, in this hypothetical case, need to secure copyright permission for the material according to the relative portion in question (factor 3).

Obtaining copyright permission can be as simple as contacting the owner of the copyright for the material, seeking written permission to use the material. However, determining who owns the copyright may become problematic. For instance, the author of a document made under hire for a client most likely transferred copyright ownership to the client as part of the original contract. Even authors of books do not typically hold copyright—that transfers to the publishers (except in cases like self-publishing). Moreover, documents produced by the US government, by law, are not restricted in use. Clearly, issues of copyright and fair use require close attention, perhaps in consultation, to avoid a lawsuit.

Evolving Responsibilities as Professionals

Just as today's world of healthcare evolves, so too do professional writers' responsibilities as professionals in the healthcare arena. Chapter 1 set the scene by discussing the responsibilities of professionals within the social contract. The material just covered in this chapter has highlighted the intertwining of legalities and ethics—in many cases, protection of rights (ethical responsibility) becomes codified (legal responsibility). Predicting further changes in roles necessitates a keen understanding of the multidimensional forces affecting the provision and delivery of healthcare. Recent US legislative changes, nevertheless, provide a hint of changes to come.

In 2001, the US Department of Health and Human Services promulgated the final form of regulations to guarantee protection of individuals' health information. Electronic availability of patient records spurred this regulation. The "Privacy Rule" was a response to enactment in 1996 of the Health Insurance Portability and Accountability Act (HIPAA). The goal of HIPAA was to simplify the electronic handling of healthcare data by common standards; these standards, however, potentially allowed misuse of private data.[8] Indirect effects of HIPAA and the Privacy Rule relate to healthcare documents with purposes such as these:

- To inform laypersons of their rights and legal recourses
- To instruct administrators on data-handling techniques
- To persuade practitioners to safeguard patients' privacy

Another legislative action with indirect implications for healthcare writers is enactment of prescription drug coverage under Part D of Medicare, the US system assisting senior citizens with healthcare costs. Individuals have been able to participate in this coverage plan since the start of 2006; however, no one is required to do so.[9] Implementation was accompanied by a degree of confusion among seniors, many requiring assistance (from the agency as well as family and friends) in understanding the plan's many options. Healthcare writers can assist seniors through preparing documents with purposes like these:

- To explain to laypersons the options and costs with the plan
- To coach administrators on responding to seniors' concerns
- To inform practitioners of revisions for therapeutic options

For these two legislative initiatives, a common thread is an increasing role for healthcare writers as patient advocates. While writing must always remain accurate and unbiased, we writers may find ourselves dealing with delicate issues, including those just discussed—information privacy and prescription coverage. Earlier in this book, we raised issues related to writing in plain language so that individuals, especially laypersons, can understand this vital information. We addressed concern for individual levels of literacy—both general and healthcare-specific literacy. Even the multicultural nature of society requires us to consider adjustments of documents for sensitivity and perhaps translation. Electronic documents raise more issues, such as whether impoverished groups without computer access are being marginalized from healthcare.

In opening this book (Chapter 1), we imagined a scenario in which we were strangers in a strange land who needed healthcare but could not speak or understand the language. Isn't that the bottom line for our responsibilities as professionals? Anyone who needs healthcare has a right to receive relevant information in a clear, concise, and unbiased manner. Through honoring ethical responsibilities, healthcare writers can deliver clear, unbiased information to today's healthcare audiences—and tomorrow's....

Chapter Summary

- As professionals entering into a social contract, healthcare writers need to act in the best interests of their audiences. While some responsibilities become codified as legalities or regulations, other responsibilities are ethically binding nonetheless. Healthcare writers must stay informed about all such responsibilities.
- Legalities for healthcare writers entail governmental regulations, such as FDA guidelines for drug development; respect for patient confidentiality and informed consent for details that may be used in documents; and adherence to rules for copyright and fair use.
- The evolution of the professional responsibilities for healthcare writers will undoubtedly introduce changes for healthcare writers. Recent legislative changes suggest that healthcare writers may take on greater roles as health advocates.

Exercises for Practice

1. Discuss the ethical implications for healthcare writers in each of the following scenarios:
   a. A pharmaceutical company hired you to prepare advertising materials on the benefits of a new drug treatment for eczema. Information provided by the client indicates little difference in the effectiveness of their new topical agent over existing lotions already available inexpensively. This is your first

job with this client, so you want to make a good impression without making any waves.

b. The insurance firm where you work assigned you a pamphlet summarizing coverage options within your plans. Your boss insists on using the firm's internal documents as source material, ideally lifting the language verbatim as much as possible. She does not allot sufficient time for you to tailor the language for laypersons.

c. You work for a medical-device group that recently developed a stent to hold open weakened blood vessels near the heart. All testing indicates that this device works extremely well, with no safety problems. Although the clinical division has not yet tested the stent in anyone over 65, the product manager pressures you to downplay the need for more research in this older population who are likely to be candidates.

d. The draft version of your cover story on AIDS among homosexual men opens with a vignette about a young man afflicted with this disease. Although open with his family and friends, he does not want his condition shared publicly; hence, he asks you to alter several facts on his background to conceal his identity.

2. Using source materials as references requires respect for the legalities of copyright and ownership. Still, the four provisions of fair use provide some leeway, provided credit is still given. For each of the following scenarios, determine how you would handle the source: no explicit credit; paraphrase and credit; quotation and credit; or credit but only with permission.

a. You need to provide the exact wording of Medicaid coverage for persons with renal failure in a pamphlet handed out in clinics.

b. An individual who overcame a severe disability and now plays paraplegic rugby gave you a fascinating interview.

c. A medical textbook provides a description of rheumatoid arthritis that you feel comfortable putting into your own words.

d. A research report contains statistics on the probabilities of outcomes for three treatments, only one of which your insurance firm covers.

e. A teenager facing brain surgery wrote a poem that would provide a solid closing for your article on healing through the arts.

3. Select a not-for-profit agency, such as a cancer society, that provides healthcare information to individuals. Search for recently enacted or pending legislation (federal, state, or local) with a potential to affect that agency's advocacy work. Come up with at least three types of documents that a healthcare writer might prepare for the

agency; be sure to include input conditions and output parameters for each. Then, consider contacting the agency with these ideas to assist their healthcare mission.

References

1 US Food and Drug Administration. *Federal Food, Drug, and Cosmetic Act* (amended). Bethesda, MD: US Food and Drug Administration; current January 5, 2010. http://www.fda.gov/opacom/laws/fdcact/fdctoc.htm.

2 US Food and Drug Administration. *Brief Summary: Disclosing Risk Information in Consumer-Directed Print Advertisements* (draft). *Guidance for Industry*. Bethesda, MD: US Food and Drug Administration; January 2004. http://www.fda.gov/Drugs/GuidanceComplianceRegulatoryInformation/Guidances.

3 US Food and Drug Administration. *Promoting Medical Products in a Changing Healthcare Environment: I. Medical Product Promotion by Healthcare Organizations or Pharmacy Benefits Management Companies (PBMs)*. *Guidance for Industry* (draft). Bethesda, MD: US Food and Drug Administration; December 16, 1997. http://www.fda.gov/Drugs/GuidanceComplianceRegulatoryInformation/Guidances.

4 American Medical Writers Association. *Code of Ethics*. 3rd rev. Rockville, MD: AMWA; updated June 2008. http://www.amwa.org/amwa_ethics.

5 International Committee of Medical Journal Editors. *Recommendations for the Conduct, Reporting, Editing and Publication of Scholarly Work in Medical Journals* (ICMJE Recommendations). Philadelphia: ICMJE Secretariat, American College of Physicians; 2013. http://www.icmje.org/recommendations.

6 US Copyright Office. *Copyright Basics* (rev). Washington, DC: US Library of Congress; May 2012. Circular 1. http://www.copyright.gov/circs/circ1.pdf.

7 US Copyright Office. Jul. 2006 (revision). *Fair Use* (rev). Washington, DC: US Library of Congress; reviewed June 2012. Document FL-102. http://www.copyright.gov/fls/fl102.html.

8 US Department of Health and Human Services. *General Overview of Standards for Privacy of Individually Identifiable Health Information* (rev). Washington, DC: US Department of Health and Human Services; April 3, 2003. 45 CFR Part 160 and Subparts A and E of Part 164. http://www.hhs.gov/ocr/privacy/hipaa/understanding/index.html.

9 Centers for Medicare and Medicaid Services. *Your Guide to Medicare Prescription Drug Coverage*. [CMS Product No. 11109]. Baltimore: US Department of Health and Human Services; revised March 2012. http://www.medicare.gov/Publications.

INDEX

From the Publisher

A name never says it all, but the word "Broadview" expresses a good deal of the philosophy behind our company. We are open to a broad range of academic approaches and political viewpoints. We pay attention to the broad impact book publishing and book printing has in the wider world; for some years now we have used 100% recycled paper for most titles. Our publishing program is internationally oriented and broad-ranging. Our individual titles often appeal to a broad readership too; many are of interest as much to general readers as to academics and students.

Founded in 1985, Broadview remains a fully independent company owned by its shareholders—not an imprint or subsidiary of a larger multinational.

For the most accurate information on our books (including information on pricing, editions, and formats) please visit our website at www.broadviewpress.com. Our print books and ebooks are also available for sale on our site.

broadview press
www.broadviewpress.com

This book is made of paper from well-managed FSC® - certified forests, recycled materials, and other controlled sources.

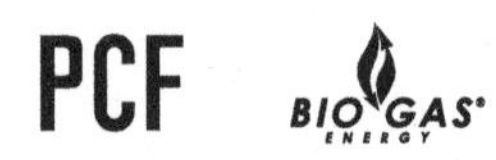